Synthesis Lectures on Professionalism and Career Advancement for Scientists and Engineers

Series Editor
Charles X. Ling, London, ON, Canada
Qiang Yang, WeBank, Shenzhen, China

This series includes short publications that help students, young researchers, and faculty become successful in their research careers. Topics include those that help with career advancement, such as writing grant proposals; presenting papers at conferences and in journals; social networking and giving better presentations; securing a research grant and contract; starting a company, and getting a Master's or Ph.D. degree. In addition, the series publishes Lectures that help new researchers and administrators to do their jobs well, such as how to teach and mentor, how to encourage gender diversity, and communication.

Ron Baecker

Ethical Tech Startup Guide

 Springer

Ron Baecker
Department of Computer Science
University of Toronto
Toronto, ON, Canada

ISSN 2329-5058   ISSN 2329-5066   (electronic)
Synthesis Lectures on Professionalism and Career Advancement for Scientists and Engineers
ISBN 978-3-031-18779-7   ISBN 978-3-031-18780-3   (eBook)
https://doi.org/10.1007/978-3-031-18780-3

This Springer imprint is published by the registered company Springer Nature Switzerland AG
The registered company address is: Gewerbestrasse 11, 6330 Cham, Switzerland

The companies that are successful ... start out to make meaning ... not money.
Guy Kawasaki[1]

And if you have a tiny startup, I have good news for you. Now is the moment you can take the most daring leaps of your career. Dream big. And act small. Pay passionate attention to your users. Handcraft the core service for them. Create a magical experience. And then figure out what part of that magical handcrafted thing can scale.

Reid Hoffman (2017), in an interview of Brian Chesky.

---

[1]https://quotefancy.com/guy-kawasaki-quotes

Acknowledgements

Thanks so much for assistance to Erik Allebest, Alex Backer, John Baker, Cristin Barghiel, Tara Currier, Karen Donoghue, Richard Hamel, James Hyatt, Adriana Ieraci, Liam Kaufman, Gordon Keller, Maureen Naughton, Lyssa Koton Neel, Mark Ruddock, Frank Rudzicz, Callum Sharrock, Philip Stern, Michael Tilson, Pooja Viswanathan, Patrick Vuscan, and Dwight Wainman for comments and ideas.

Kim Davidson, Andrew Silverman, and John Unsworth have been especially supportive with comments and suggestions.

Vincent Pham assisted me throughout the writing process in many ways and always with skill and diligence.

Ian Small deserves a very special thank you. Taking advantage of his 30 years of mostly-Silicon-Valley experience in innovation, entrepreneurship, leadership, and management, he critically and constructively read my manuscript in detail, wrote pages of critique and suggestions, and continued to help with numerous phone and email chats. This book is MUCH better thanks to Ian.

All errors and omissions and misunderstandings and misstatements are of course solely my responsibility.

Introduction

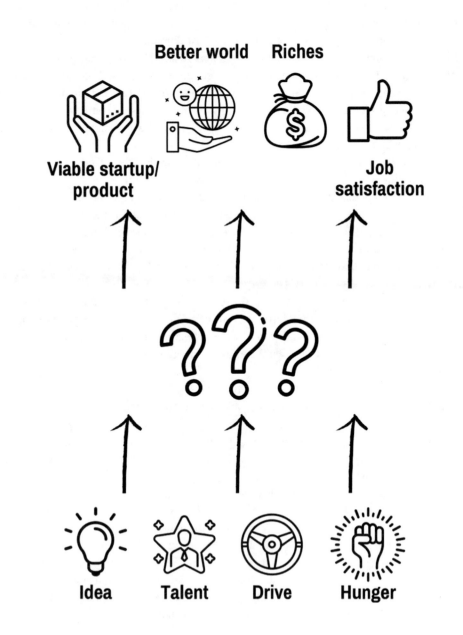

I am the unlikeliest of entrepreneurs. I am the child of Holocaust survivors. My father spent four months in Dachau when it was still possible to get out. My mother was sent from Vienna to England by her parents and never saw them again.

Yet a few months after getting tenure at the University of Toronto, on January 1, 1976, I became a co-founder and CEO of my first software enterprise, Human Computing Resources Corporation, HCR for short, which we spun out of our Dynamic Graphics Project (DGP) laboratory. HCR's full name was chosen to suggest an ethical stance, as we planned to apply computer science for the good of humanity.

Since then, I have created or been instrumental in the founding and growth of five software companies, two publishing ventures, three university research institutes and laboratories, a Canada-wide research network, and a non-profit virtual community. I started teaching University of Toronto students how to start and grow successful software enterprises in 1986, and since then, I have taught such courses to thousands of students on three continents.

This guide summarizes what I have learned in almost 5 decades of being an entrepreneur. Why might you want to become an entrepreneur?

The first reason is that there is some activity you want to do or some artifact you want to build. For accomplishing something, you may want to create a services company, a university laboratory, or a non-profit organization. To create something, you may want to start a software products or digital media firm. In all cases, you must have a guiding vision, a dream of something that is better—substantially better—than what exists currently.

The second reason is that you want to use technology to make the world a better place. Technology, according to Merriam-Webster's Dictionary, is "the practical application of knowledge". I take a very broad view of technology, including far more than digital technologies—not only software and digital media, but tech for health, learning, and a sustainable environment.

I use the term "ethical tech startup" in two ways. These correspond to two ways of parsing the phrase—"ethical tech" startups and ethical "tech startups". Many of the case studies are firms commercializing ethical technology—products and processes to support health, learning, community, recreation, productivity, and a sustainable environment. My cases are also tech startups trying to behave ethically, sometimes making moral decisions in the face of hard choices.

I have been asked by a thoughtful friend what I mean by "ethical behavior"? Three credible online descriptions of ethical behavior in business are "knowing right from wrong", "acting in a manner that is in tandem with what society considers to be good morals", and "doing the right thing and adhering to professional standards". My definition is consistent with these, incorporating both knowledge and action, and guided both by an abstract concept of what is good and by societal and professional standards.

The third reason for becoming an entrepreneur is that your personality, instincts, and skills make you better suited to running an enterprise than working for somebody else.

You may feel this after several dismal or even brutal experiences reporting to people who lacked intelligence or kindness or were just limited in their vision.

The final reason for being an entrepreneur is that you want to make a lot of money. As of 2022, 7 of the world's 10 wealthiest billionaires made their money in technology. Microsoft alone created over 10,000 millionaires. Unicorns are privately held startup companies worth over $1 billion. There are now approximately 1000 unicorns. Of the 10 with the greatest valuation, 8 are tech firms. The opportunity to become wealthy does not seem to reduce over time; in other words, we are not running out of new ideas or opportunities.

Let's be clear. Most of what I shall talk about in this book has nothing to do with making a fortune. What it is always about is taking a vision and turning it into a reality and about structuring growth so that the venture is sustainable, so that it can continue to achieve and prosper over time.

Clearly, there are many compelling reasons for being an entrepreneur. But why did I write this book? Primarily, it is because I have learned much over these almost five decades that I wanted to share with readers. I started writing it longhand while in a hammock in Provence in the spring of 1985 under the title "The Business of Software". Although I then taught a course by this name for the next 25 years, I did not have time to restart the book.

In rewriting my course notes in the spring of 2020, I realized that I now had a much deeper understanding, achieved by teaching courses over 35 years, speaking with hundreds of entrepreneurs, reading avidly, observing firms do well and make mistakes, starting and growing my own ventures, and advising many ventures.

I therefore wanted to share my insights with others, so they could succeed as ethical entrepreneurs. What follows in the rest of this book is my understanding, delivered in the form of a clear and concise presentation of principles for ethical tech startup success. The recommendations are grounded in the experiences of 26 example firms and the entrepreneurs whose leadership and skill animated them.

The book is organized into nine chapters.

Chapter 1 deals with value. Value for customers is essential for startup success. Entrepreneurs need to have visions of something substantially better than what now exists. The process of innovation articulates such dreams; entrepreneurship turns innovations into sustainable ventures. Such ventures need be responses to societal problems and technological opportunities that enable entrepreneurship as a firm develops innovative solutions to the problems. Solutions can be delivered to markets consisting of customers. Targeted niche markets are highly desirable for most startups. I introduce the concept of a value proposition, a method of describing in qualitative and ideally quantitative terms the value a solution to a problem brings to a firm's customers.

Chapter 2 deals with achieving a competitive advantage over your competition. (There is always competition.) The advantage typically comes from some underlying magic, the special sauce that your venture delivers to provide a solution to a problem.

The most powerful magic is intellectual property (IP), often consisting of ideas, inventions, processes, or algorithms. I describe major methods to protect IP—trade secrets, copyright, patents, and trademarks—and how IP may be protected. I also discuss other concepts important to achieving competitive advantage—distinctive competence, first mover advantage, and barriers to entry.

Chapter 3 deals with validation, methods of ensuring that the problem is important and that your solution is viable and compelling. You must identify relevant stakeholders and create personas of your target customers, elicit customer insights into the problem and the solution, and develop and test use cases for your solution. User experience research will help you understand if your solution can be effective. Methods to achieve insight include ethnographic research, building and refining product prototypes, usability testing, and responsive and thoughtful customer support. Validation also happens through competitive analysis, a disciplined, honest comparison of what you offer relative to competition. Such analyses will force you to articulate your differentiation, how you will stand out favorably in a crowd of competitors. Substantial validation can be done before investing hundreds of thousands of dollars in product development.

Chapter 4 deals with statements of identity. It is essential that you know who you are, what you are good at, and why you exist. Ways of describing the essence of a venture include vision statements, mission statements, company and product names, mantras, and logos, all of which help define a brand. These assist you in developing assertive self-descriptions, such as elevator pitches—pithy single sentences of a firm's vision, mission, and excellence—and pitch decks—10-15 slide deck presentations to convince potential investors to advance funds to an enterprise, new employees to join, customer prospects to buy, or media to believe that you have created something very special.

Chapter 5 deals with the encapsulation of all the above into a business model, a structured way of visualizing the essence of an envisioned enterprise. I shall present the useful abstraction of a business model canvas. Key aspects are a revenue model, defining how you will earn revenue and be profitable, and a go-to-market plan, describing how you will inititally introduce a product or service into a market. I shall discuss the 4Ps of modern marketing—product, price, place, and promotion—emphasizing defining the product, modern pricing of apps, strategic design of a web and social media presence, and internet marketing. I will stress the importance of entering a market quickly with a minimal viable product and the value of customer relationships in sustaining viability and achieving competitive advantage.

Chapter 6 deals with finances. I shall explain the fundamentals of financial statements and the key performance indicators which help you to monitor the financial health of your venture. I stress the importance of a financial forecasting model for managing an enterprise and attracting financing. In an appendix, I shall explain a rubric for a concise spreadsheet that I developed and have validated in my teaching and consulting. (It is freely available on the website https://ronbaecker.com/ethical-tech-startup-guide/tech-corp-forecasting-rubric/.). I shall discuss the concepts of debt and equity and introduce major methods of financing, including borrowing money, applying for grants, and selling stock, with methods including sweat equity, family and friends financing, angel investments, venture capital, and going public. I discuss the importance of having both growth and exit strategies for a startup.

Chapter 7 deals with the people side of the equation, the team that forms the heart and soul of an enterprise. This is arguably the most important. I shall distinguish between leadership and management and stress the importance of a visionary and humane leader, a resourceful management team, and employees with incredible talent and commitment. I shall place special emphasis on the concept of startup culture; creating a vigorous and collaborative culture is one of a leader's major tasks. I shall discuss the courage and creativity it takes to orchestrate a strategy pivot when the company's progress is stalled. I shall also speak of the roles of a wise and independent Board of Directors or Advisors and of strategic partnerships that include alliances with firms that are non-competitive and even some that are competitive, distributors, universities, and the media.

I discuss corporate ethics throughout the book, reviewing ethical challenges in the context of each chapter. I stress that the leaders of an enterprise have ethical responsibilities to their employees, customers, shareholders, and the public. I bring this all together in Chap. 8 with a discussion of substantive issues and ethical choices in specific cases, some of them controversial. Examples of ethical dilemmas and ethical and non-ethical choices and actions are presented, including situations in the life of well-known firms such as Apple, Microsoft, Airbnb, Twitter, and Uber.

Chapter 9 extracts concise statements of wisdom from the book as sixty Principles for Success for ethical tech entrepreneurs.

Throughout, I illustrate principles by describing specifics using 26 case studies. These are listed after the Table of Contents and are described in 1-page descriptions interspersed among the chapters. Most of these are software firms, but some develop both hardware and software, and some aggregate content. There are examples of "ethical technology" firms in the health, learning, community, recreation, and environmental sustainability marketplaces. All have governed themselves ethically, but I shall highlight in Chap. 8 examples of unethical behavior over major periods of time. Three of the startups began in 2021. Details about these firms have been checked in conversations with founders or by accessing information found in at least two independent reputable sources.

Who should read this book and why should they read it?

My primary audience is individuals contemplating a technology startup or already working for one. In most cases, they are technically trained individuals with idea, talent, drive, and hunger for success and with visions of creating a big company with a breakthrough product. Many such individuals will want to change the world for the better. Most will seek riches as their reward. But they typically have no business background and have no idea where to begin.

No worries—you do not need an MBA to create a viable tech startup! Reading this book will improve your ability to decide whether you should create a startup, will guide you in doing so knowledgeably, professionally, creatively, and ethically, and then steer you wisely toward growth and profitability.

And read it several times … before starting your venture, and while you are underway. It will also inform individuals joining a tech startup as a manager or employee.

My book will be of particular interest to idealistic young people seeking to do good with technology at a time when so much of it has been subverted in damaging ways.

Another audience is the partners, children, parents, or other family members of startup participants. Reading this book will help them be more supportive partners, children, and parents.

Others who will find the book interesting and valuable are thoughtful persons trying to understand both the good and the problematic directions digital technology is taking us, and why so many tech people are so passionately committed to their work.

Entrepreneurship, if done well, still seems to be like "magic".

- It seems like magic when science has a commercializable breakthrough.
- It seems like magic when a team creates what wasn't thought possible.
- It seems like magic when new software starts working for the first time.
- It seems like magic when customers are delighted with new products.
- It seems like magic when a management team works smoothly.
- It seems like magic when entrepreneurs make the world a better place.
- It seems like magic when startups go public, creating new millionaires.

My goal in writing this book is to share this magic with you and to guide you in creating and growing your own successful ethical tech startup.

Contents

Case Study Ventures

About the Author

Ron Baecker I am Emeritus Professor of Computer Science and Bell Chair in Human-Computer Interaction (HCI) at the University of Toronto (UofT), also co-founder of UofT's Dynamic Graphics Project, and founder of both its Knowledge Media Design Institute (KMDI) and its Technologies for Aging Gracefully lab (TAGlab). I currently am an Adjunct Professor of Computer Science at Columbia University, where I teach "Computers and Society".

My B.Sc., M.Sc., and Ph.D. are from M.I.T. I am an ACM (Association for Computing Machinery) Fellow, a member of the ACM CHI (Computers and Human Interaction) Academy, one of ACM's 60 Pioneers of Computer Graphics, and a Canadian Digital Media Pioneer. I was recently awarded the 2020 ACM CHI Social Impact Award. I have written over 200 papers and articles, mentored over 200 outstanding students, and inspired thousands through my teaching.

I designed and built Genesys, the first computer animation system of significant generality (1966–9); this work helped launch the field of computer animation. My computer-animated computer science teaching film, Sorting Out Sorting (1973–81), is an underground classic, loved by students worldwide; it helped launch the field of software visualization. My most recent research focused on envisioning, designing, and evaluating technological aids for aging gracefully, intended for individuals with Alzheimer's disease, mild cognitive impairment, amnesia, vision loss, and stroke, and for normally aging senior citizens.

My work has also been recognized by leading firms developing innovative digital technologies. I was invited to spend the summer of 1974 in Alan Kay's Smalltalk group, contributing to the development of the Xerox PARC technology that inspired the Apple Macintosh and Microsoft Windows. I was also invited to spend half of 1988 in Apple Computer's Human Interface Group, a research team that helped ensure the achievement of Steve Job's vision of the Macintosh as "a computer for the rest of us".

In 1976 I co-founded and became CEO of Human Computing Resources Corp., HCR for short. I have created or been instrumental in the takeoff and growth of five software companies, two publishing ventures, three university research institutes and labs, a Canada-wide research network, and a non-profit virtual community.

A few words about my 5 startups. HCR was a world-class UNIX systems software firm, sold in 1990 to the Santa Cruz Operation. Expresto Software failed with its product for movie authoring; the company was sold in 2002 to a shareholder. Captual Technologies successfully commercialized the ePresence rich media webcasting and archiving system; it was sold in 2011 to Desire2Learn. MyVoice Inc. commercialized a context-aware mobile speech aid app for individuals with speaking challenges. Famli.net Communications is commercializing a novel app for older adults and their families and communities, especially those whose use of technology is impeded by fear, language, vision, hearing, hand tremor, or language competency challenges.

Since 1986, I have been teaching principles of software entrepreneurship at the University of Toronto, including most recently the summer of 2021. I have taught thousands of students since then at U of T and in short course form on three continents.

Three textbooks on CHI and CSCW helped launch these fields as academic disciplines. I am the founding Editor of the Synthesis Lectures on Technology and Health (Springer Nature, Publisher). In 2019, I published the first comprehensive and current textbook dealing with computers, society, and ethics—Computers and Society: Modern Perspectives (Oxford University Press). In 2020, I was the lead author of two editions of the 2020 book The COVID-19 Solutions Guide: Health, Wealth, Technology, and the Human Spirit. My 2021 book was Digital Dreams are Now Nightmares: What We Must Do. ACM Press will publish a second edition in 2023. I am also the founder of a resource hub bringing together and organizing Computers and Society literature, computers-society.org.

I speak widely on these issues—12 talks in Canada, the USA, Mexico, Great Britain, Singapore, and India in 2019 and early 2020. Talks are now resuming via Zoom, for example recent speeches in Japan, New Mexico, Pakistan, Thailand, California, and Boston; face-to-face talks will start again as I speak around the world in my new role as an ACM Distinguished Speaker. I have been interviewed widely by print, radio, video, podcast, and other internet media journalists.

A. Imax Corporation

Since the early days of motion pictures, inventors have tried to embed viewers into a more immersive experience by using larger film formats and screens. Although Cinerama, Cinemascope, and Vistavision failed in the 1950s, Canadian filmmakers Graeme Ferguson, Roman Kroitor, and Robert Kerr produced successful multi-image, multi-screen productions for Expo' 67 in Montreal. Because of this success, they were approached by Fuji Bank to produce an experimental film for Expo' 70 in Osaka, Japan. Together with engineer William C. Shaw, they accepted the contract and formed Multiscreen Corporation, later to be renamed Imax.

They were way out on a limb, facing many challenges. The first was the design of a new motion picture camera that could record on 70 mm film running horizontally instead of vertically. Far more difficult was the design of a new projector that could withstand the mechanical challenges of using the larger format. They also had cash flow issues, but happily they managed to get an advance from Fuji. They delivered the system on time, yielding a magical experience for viewers; their projector was operational for all but one day of the six months of Expo '70.

The next challenge was to move the startup from delivering isolated experiences for special customers to making the Imax experience widely available, in other words, to move from prototype to product. The Government of Ontario bought the Osaka projector and installed it at Ontario Place in Toronto in 1971. After that, they were able to demonstrate the concept could work on a domed screen in a new kind of planetarium in San Diego.

Over the intervening 50 years, the company has expanded to locations all over the world. Imax experiences are available in smaller formats which can be retrofitted to existing multiplex theaters, and 3D projection and high-resolution digital imagery. Imax productions now include documentaries as well as full Hollywood feature films. There are currently over 1600 Imax screens in over 80 countries.

B. Microsoft Corporation

In 1975, computers were either thousand-pound mainframes costing millions of dollars or hundred-pound minicomputers costing tens of thousands. Tech enthusiasts were therefore delighted at the announcement of the Altair microcomputer kit for a few hundred dollars. Harvard undergraduate Bill Gates, together with his high school tech buddy Paul Allen, worked day and night to demonstrate in New Mexico a version of the basic programming language, a solution to the problem of making the hardware useful. Basic allowed programmers and even non-programmers to create applications; the applications drove the launch of the personal computer revolution. Microsoft would soon be the dominant software firm of the 80s and 90s.

Their insight and actions illustrate three other themes of this book, the importance of an ambitious, far-seeing, and accurate vision; the first mover advantage afforded a new firm if it is the first in the market; and the leverage gained from an intimate involvement with leading-edge research and development.

Gates and Allen soon purchased software than enabled them to build and license to IBM in 1981 the MS/DOS operating system. MS/DOS was later replaced by Microsoft Windows, which fueled the personal computer revolution. Their guts and cleverness in making the deal with IBM before they had the means to deliver what they promised illustrates the importance of confidence and living on the edge that characterize many successful tech startups.

Microsoft from the start was the leading source of systems and application software for personal computers and has continued to excel, in part because it created and maintained the foremost industrial computer science research laboratory in the world. Its software has enabled enormous growth in the world's productivity and creativity. It seemed to lose its way ethically in the 90s and became known as the "Evil Empire", yet it has been re-energized and to a significant extent become a force for good under third CEO Satya Nadella.

C. Apple Inc.

Marketing and design genius Steve Jobs and technical superstar Steve Wozniak introduced the Apple 1 microcomputer kit in 1976 and the consumer-friendly Apple II in 1977. The latter attained rapid success because of its color graphics, open architecture, and software for personal use. Visicalc, the first spreadsheet, became the "killer app" which drove its use for business and helped the company achieve over $100 million of sales by 1980.

Apple's 3-day visit to Xerox PARC in 1979 in which Jobs was shown an incredible array of interactive software developments stimulated its development of the first Graphical User Interface (GUI) computer in 1984, the Macintosh. The lesson is to be on the lookout for great ideas and capitalize on them. Not all ideas have to come from within the firm, a lesson we shall encounter again.

After years of management turmoil including the departure of Jobs, Apple brought him back in 1997 and started to recover. Between then and his resignation due to health reasons in 2011, his vision and leadership drove the development of breakthrough products including the iPhone smartphone, the iPad tablet, and the Apple (digital) Watch, as well as marketing innovations such as tracks of music sold for 99 cents from iTunes, the App Store, and elegant retail stores. Success was also driven by effective and graceful graphic, industrial, and media design, speaking of quality and usability.

Jobs was not afraid of breaking the rules. For example, it seemed outrageous at that time to create high-end retail outlets for the products of a single tech firm, but his instincts on this front were right on and helped drive consumer confidence and adoption.

Apple's focus on usability enabled a vast expansion in individuals without technical training using technology comfortably and safely. Yet recent concerns about tax avoidance, commissions that seem unreasonably high, and unfair competitive practices have raised legitimate concerns that the firm is no longer behaving ethically, a topic I shall discuss further in Chap. 8.

Value

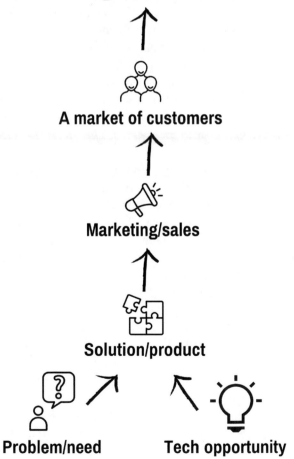

Problem mitigated / need addressed

A market of customers

Marketing/sales

Solution/product

Problem/need   **Tech opportunity**

R. Baecker, *Ethical Tech Startup Guide*, Synthesis Lectures on
Professionalism and Career Advancement for Scientists and Engineers,
https://doi.org/10.1007/978-3-031-18780-3_1

Seven Sources for Innovative Opportunity

1. *The unexpected*
2. *Incongruities*
3. *Process need*
4. *Industry and market structures*
5. *Demographics*
6. *Changes in perception*
7. *New knowledge.*

Peter F. Drucker (1985). *Innovation and Entrepreneurship: Practice and Principles.* Harper & Row.

In the fall of 2015, at the urging of a senior medical research scientist and a technology entrepreneurship expert, a partner from the venture capital firm Versant had a late afternoon drink with Dr. Gordon Keller, Director of the Stem Cell Institute at the Princess Margaret Cancer Centre in Toronto. By then, it was well established that stem cells were an exciting new technology that could be harvested and "reprogrammed" to behave in specialized ways. Versant saw this as an opportunity to address classes of medical problems such as heart attacks. The latter idea arose because Dr. Keller's lab was active in creating heart cells that could plausibly be used to repair heart tissue damage resulting from attacks. To turn this innovation into a solution for significant medical problem, Versant together with the pharmaceutical giant Bayer AG created the entrepreneurial firm, Blue Rock Therapeutics.

This chapter explores how technology innovation directed at societal problems and exploiting technological opportunities can serve as the foundation for entrepreneurial ventures offering novel solutions to these problems. I shall introduce the concepts of *markets* consisting of large groups of potential customers and *niche markets* comprised of smaller groups which are ideal for startups. If a niche market based on an important societal problem or need is well chosen, and customers are obtained through effective sales and marketing (discussed in Chap. 5), then a startup can provide value to its customers and achieve viability, growth, and profitability.

Innovation and Entrepreneurship

Innovation, according to the Merriam-Webster Dictionary, is "the act or process of introducing new ideas, processes, or methods". The Business Development Bank of Canada notes that "innovation is really about responding to change in a creative way. It's about generating new ideas, conducting R&D, improving processes or revamping products and services." Yet neither of these two formulations discuss how to take what is new and deliver it effectively to customers.

That is where the concept of *entrepreneurship* comes in. The website oberlo.com defines entrepreneurship as "the act of creating a business or businesses while building and scaling it to generate a profit. ... The more modern entrepreneurship definition is also about transforming the world by solving big problems." Most entrepreneurial ventures are for profit, but there are many non-profit entrepreneurial ventures.

It is also possible to create a small new venture within a larger existing business. This is called *intrapreneurship*; its principles are like those of entrepreneurship, except it is easier, as the host company may provide resources; more complicated, because of large firm politics; and less financially rewarding, because you are doing the work as a salaried employee. Yet, if your primary goal is to make a better world, do not overlook the opportunity to do it from within an established firm with a track record and a path to market and with the cash to bring a new product to market.

Ethical Missions

Successful innovation is a response to perceived *opportunities* to do things better or to solve important *problems*. Examples of opportunities are global, near-universal connections via the internet; stem cell harvesting being practical; and increased awareness of the environmental crisis and the urgent need for responses. Examples of societal problems are too much carbon in the atmosphere, global pandemics, shelling of civilians during warfare, and an inability to concentrate while in the classroom.

Antibe Therapeutics, Winterlight Labs, Blue Rock Therapeutics, and Braze Mobility are devoted to improving our health and fighting disabling or deadly disease. Desire2Learn, the New York Times, Wikipedia, and MasterClass support learning and knowledge acquisition. Shaw Industries, Ecovative Design, and Beyond Meat provide products sensitive to environmental concerns. Apple, Microsoft, Adobe, Caseware, LinkedIn, Twitter, Canva, and Nuula have expanded the power of individuals and businesses through access to computing and communications. Imax, SideFX, Chess.com, Airbnb, QLess, Drisit, and Wordle support leisure, travel, entertainment, and convenience.

Note that, except for Blue Rock Therapeutics, I am not speaking of moonshots, companies trying to solve "grand challenges" like a cure for cancer or an inexpensive large-capacity rechargeable battery. I am simply suggesting that there are many opportunities to make the world a better place, and that having an ethical mission will be good for an entrepreneur's soul and will attract young people with both excellence and a social conscience to join the team.

Opportunities for ethical entrepreneurship range from the glamourous and high-profile to the quietly impactful. Some may not be sexy, as is the case of Shaw Industries, a good example of a firm moving steadily and consistently towards a noble aim. Since reinventing itself with a commitment to environmental sustainability in 1999, Shaw is now proud that 90% of its manufactured products are Cradle to Cradle Certified®.

Opportunities and Problems

Here are some examples of technological opportunities being exploited by our case study firms:

Imax: People love movies immersed in a new reality. Movies would be more compelling if they could be made more immersive. Recording on and projecting from larger formal film stock could make this possible.

Microsoft: The advent of the personal computer marketplace generated a huge opportunity for software, beginning with programming languages (Basic), operating systems (MS/DOS), and productivity tools (Word).

Wikipedia: The success of the world wide web and tools for non-experts to create web pages using wiki software provided a way for thoughtful individuals worldwide to build a new kind of encyclopedia.

Adobe: The explosive use of personal computers and the availability of digital printers created the opportunity for widespread digital document creation and processing.

Twitter: Universal availability of messaging allowed creation of a novel social medium appropriate for announcements and transmission of bursts of information.

Chess.com: The increasing use of digital devices and the long-standing interest in the game of chess resulted in an opportunity for a high-quality app for playing, learning, and improving one's game.

Airbnb: Digital matching allowed homeowners to earn income by short-term renting of extra rooms in their home and allowed travelers more flexibility and congeniality in finding accommodations while traveling.

Beyond Meat: Awareness of cruelty to livestock and the contribution of meat to the environmental crisis created opportunities for "meat" made from vegetables if the products could be sufficiently appetizing.

Winterlight Labs: Improvements in speech recognition allowed its successful use in detecting cognitive decline.

Drisit: The widespread availability of drones generated new opportunities in tourism and education and for utilities and families.

Here are some problems being addressed by our companies:

Apple: Traditional digital technology was hard to learn and use by all those without technical training.

NYTimes.com: The availability of news and advertising on the web was dramatically reducing newspaper advertising revenues, making them financially unviable, and killing many of them.

LinkedIn: Obstacles to effective work-oriented global social networking made it diffi-cult to make new business acquaintances and in so doing to improve one's career and job prospects.

Antibe Therapeutics: Severe pain affects hundreds of millions of people worldwide; many medications for pain are dangerously addictive or damage the digestive tract.

QLess: Great frustration has been experienced while waiting in long lines to get into places such as amusement parks, government and university offices, and hospital departments.

Ecovative Design: Packing materials are used in enormous quantities but are not biode-gradable, posing problems for the health of wildlife, the appearance and tranquility of the outdoors, and the future of the planet.

Canva: Existing tools to create graphic designs such as social media posts, brochures, and posters were too difficult to learn and use and too costly.

Braze Mobility: Mobility-challenged people with cognitive impairments and sensory disabilities were being denied use of wheelchairs in institutions due to safety concerns.

Nuala: Small business owners have lacked actionable financial data and ready access to emergency capital.

Another way of thinking about problems that can be solved by technology is the con-cept of *pain points*. Standing in long lines at an amusement park is a source of pain for families with energetic children; it also causes pain for proprietors of shops whose poten-tial customers are in the lines and not in their shops. Pain medications that are addictive are "pain points" for people suffering from chronic or acute pain. Lying immobile in a hospital bed for weeks is a pain point for individuals forced into an intensive care unit (ICU) by COVID-19. An impetus for the PhD research of Braze Mobility Founder Pooja Viswanathan was that older adults with dementia were being denied the use of wheel-chairs in long-term care facilities.

Pain can be understood qualitatively, as for example the stress of parents trying to calm and distract bored children. Even better is a quantitative measure, such as potential revenue per day in amusement park shops that was lost, or the cost to society caring for people who have been incapacitated by a disease, or the number of deaths attributable it.

New ventures need to seek solutions that mitigate or remove pain.

Successful startups exploit technological opportunities to provide solutions to known problems. Here is a summary of the opportunities and problems for our 26 case study companies:

Startup	Opportunity	Problem
A. Imax	Large format film	Movie screens too small and flat
B. Microsoft	Software for PCs	PCs exploding, need software
C. Apple Inc	Graphical user interface (GUI)	Computers hard to learn and use
D. Adobe Inc	Publishing, graphics software	Desktop publishing s'ware need
E. SideFX	Procedural comp. animation	Hard to create complex scenes
F. Caseware Int'l	Interactive s'ware applications	Auditing slow, $$$, error-prone
G. NYTimes new	Internet, world wide web	Newspaper viability threatened
H. Shaw Ind. new	Sustainable floor coverings	Environmental collapse danger
I. Desire2Learn	Internet, web	Learning management complex
J. Wikipedia	Web, wikis	Knowledge huge, inaccessible
K. LinkedIn	Web	Job search, networking hard
L. Antibe	Medical hydrogen sulfide	Addiction or stomach damage
M. Twitter	Internet messaging	Global info broadcasting need
N. QLess	Mobile queue management	Standing in long lines
O. Ecovative design	Mushroom binding materials	Non-biodegradable packaging
P. Chess.com	Internet, chess apps	Online chess resource need

Startup	Opportunity	Problem
Q. Airbnb	Internet, 2 sided apps	Hotel rooms too expensive, impersonal
R. Beyond meat	Plants can taste like meat	Meat damaging to planet, health
S. Canva	Software as a service	Graphic designs hard to create
T. MasterClass	Internet video streaming	Desire to learn from the best
U. Winterlight Labs	AI speech processing	Monitoring dementia need
V. Blue rock	Stem cell research	Safe tissues for surgery need
W. Braze mobility	AI sensing of surround	Wheelchair navigation need
X. Drisit	Ecology of drones	Desire for seeing at a distance
Y. Nuula	Apps for small business	Timely info and cash need
Z. Wordle	Online games and puzzles	Excellent 5-min word puzzle

Successful tech startup ventures develop elegant solutions that solve problems by exploting technological opportunities.

Solutions

Here are some examples of solutions delivered by our case study firms:

Adobe: From a modest start with the creation of Postscript to support document creation, publishing, and printing, Adobe has added through research and shrewd acquisitions solutions for allied markets such as Photoshop for graphics editing, Premiere for video creation and editing, and Pagemaker for page layout.

LinkedIn: The firm invented a novel social medium catering to the needs and style of professionals eager to network online for the purposes of career advancement, job search, and learning.

Antibe Therapeutics: This company is developing and clinically testing non-addictive oral painkillers and anti-inflammatories that are based on its hydrogen sulfide releasing platform.

Twitter: The firm invented a novel digital medium consisting of text "sound bites", useful for business, marketing, information dissemination, and social communication.

QLess: Its solution consists of an app allowing people about to enter huge lines with long waiting times to register in virtual queues, giving them the freedom to shop or explore or dine until they get text messages asking them to return and join the physical lines for the last few minutes.

Beyond Meat has a solution for people requiring the apparent taste and sensation of eating meat but who no longer want to eat meet because of health or environmental concerns.

Canva: Its solution for people wishing to create their own digital graphic designs is software that is easy for people to learn and use.

Winterlight Labs created an app that monitors cognitive decline unobtrusively through the analysis of speech and that has proved attractive to pharmaceutical firms developing medications relevant to dementia care.

Blue Rock Therapeutics is developing ways to grow heart cells in a lab, which can be used to improve the effectiveness of heart transplants.

Braze Mobility focused its expertise in intelligent wheelchairs by developing smart sensors to solve the specific problem of aiding navigation and avoiding collisions with the wheelchair.

I discuss how to turn solution ideas into products in Chap. 5.

Markets and Market Segments

You must be careful. An elegant solution to an unimportant problem may be of little value, unworthy of an entrepreneurial effort. The problem must be a big one and the market large enough to warrant the time, energy, and investment capital to justify starting a new venture with the potential to achieve significant success and business value.

A *market* is a set of possible customers. We speak of consumer markets, such as all homeowners in the U.S., all the world's blind people, or all Canadian children between the ages of 5 and 11. Many tech businesses are labeled as B2B or B2C depending upon whether they address business or consumer markets. B2B firms may sell to all businesses in countries where French is spoken, all U.S. law firms, or all companies packing and shipping more than 10 million boxes per year. B2C firms may sell to all European persons with annual earnings of over 100,000 Euros, all Canadian women who are senior citizens, or all high-income retired people interested in continuing education. I shall return to this topic in Chap. 5.

Markets are comprised of subsets of customers known as *market segments*. For example, Imax addresses niche markets such as large multiplex movie theatres. Adobe addresses niche markets such as graphic design, illustration, and film editing. Desire2Learn addresses niche markets such as universities, colleges, and secondary schools; it also segments its markets geographically. QLess addresses industries that represent subsets of locales where people stand in lines, e.g., amusement parks, government agencies, universities, and hospitals. Winterlight Labs c currently addresses the niche markets of pharmaceutical companies; it might at some point also focus on dementia care institutions. Braze Mobility could address smart wheelchair needs among subsets of wheelchair users such as those with impaired mobility, vision, hearing, or cognition, or users in different settings, such as retirement homes, long-term care facilities, or hospitals.

Niche Markets and Focus

Start-ups are wise to focus initially on "small" segments that are called niche markets. A *niche market* is one that is small enough to afford a new firm a good chance to attain a dominant position if the solution is excellent and unique and the firm is well managed. Reasons the market must be "small enough" are to enable reaching customers at reasonable cost and so that larger predatory firms have little interest in entering the market and squashing the startup. Choosing an initial niche market in no ways restricts your vision and ambitions, which can be huge (see below).

Here are some niche markets addressed by our case study firms.

SideFX: This innovative small software developer addresses four market segments: film and television, game development, motion graphics, and virtual and augmented reality. Its technology is ideally suited for a niche within these markets—motion pictures and digital media needing complex special effects involving thousands of similar-looking and -behaving objects, such as crowds, forests, waves, or clouds.

Caseware International: Auditors in carefully selected countries, but not the much larger market of accountants, and initially not including the U.S. as a target geography.

LinkedIn: Professionals wishing to improve their career success, ignoring the much larger general social media market.

QLess: Institutions in specific markets such as government, education, health care, sporting events, banks, and retail.

Chess.com: People interested in chess, ignoring much larger markets such as those interested in poker or fantasy sports.

Winterlight Labs: Pharmaceutical firms, which are the institutions with the deepest pockets among the many institutions and individuals concerned with cognitive decline.

Wordle: Word-game enthusiasts who are online daily and who appreciate a challenging game that only takes 5 min.

There are many other reasons why a *focus* on a niche market is advantageous for a startup. Money for marketing is usually tight; focus enables smaller budgets to be targeted at specific segments. Focused participation on a segment gives you greater visibility and traction with media, influencers, and customers. Management also can direct its attention and not be distracted with multiple conflicting priorities.

Focus applies to both products and markets. Startups should begin with one product and market and concentrate on establishing its excellence and maintaining its superiority before becoming distracted with follow-on products or markets. Firms striving for growth and massive success (as opposed to *lifestyle businesses* which are content to achieve and remain at modest size and sufficient profitability for the needs of its founders) will anticipate at the beginning and refine over time methods for growth, either via adding new products or new markets or both.

Company founders are often reluctant to embrace niche markets because of their relatively small size and because they fear that they will be trapped in a small sandbox forever. This fear is groundless.

If successful in an initial small niche, they will have growing opportunities to develop enhancements to their core product which can be independently monetized, as well as to leverage their position in a market with innovative follow-on products. More importantly, if they attain a commanding position in one market and have good reference customers attesting to product quality, they can enter new markets. The business consultant and author Geoffrey Moore describes this as a "bowling pin" strategy for company growth.

Imax: Imax's first customers were an exposition in Japan and a theme park in its home geography of Canada. Over the past five decades, it has expanded relentlessly to over 80 countries and has enlarged its product offerings to suit a variety of theatre sizes and shapes.

Caseware International (a Toronto-based firm): Caseware began with one geography—Canada; after achieving a commanding position in Canada, it moved on to one European country; as it achieved momentum and credibility in each geography, it continued entering attractive new markets in various parts of the world; it then entered the U.S. when its financial position was solid enough to sustain entry into this very large and competitive market.

QLess: Being uncertain as to which of its niche markets would be most tractable, Qless attacked sectors such as government, education, retail, and health care, achieving degrees of success in all of them.

Airbnb: Airbnb expanded geographically relentlessly after product launch and traction in the U.S. It emphasized room rentals in private homes at the beginning, then quickly also supported rentals of apartments and entire homes. It now offers "experiences" as well as places to stay.

Beyond Meat: It expanded both in terms of products, such as artificial chicken, beef, and pork, and in terms of distribution in countries all over the globe.

Such strategies assume that the individuals or businesses targeted attend the same conferences or read the same magazines or belong to the same trade organizations, which allow cost-effective promotion to all similar sales targets.

Saying "NO"

Focus is key. *Saying "no"* to other opportunities will allow you to develop a strong market position in your niche market from which you can later expand geographically and/ or with follow-on products. It will enable your management team to concentrate on what is most essential, avoiding distractions. It will also help keep your staff lean, reduce your capital needs, and convey a clear identity and message to customers, customer prospects, and the media.

Yet the most important reason is to protect the dominant position of your firm in its chosen niche. Here are three examples of firms that did not do this and failed as a result.

Human Computing Resources Corp. (HCR), my first startup, achieved a position by 1981 as one of the 4 leading UNIX systems software firms in the world, focused on porting UNIX software development tools. My successor as HCR's CEO in 1984 did not appreciate this, and almost bankrupted the company as he tried to move into developing and selling business applications software, ignoring the fact that we lacked this competence and could not excel in the business software domain.

In 1978, Dan Bricklin had the idea of a calculator driven by a mouse that could enable easy computations on a display of rows and columns. Bricklin and Bob Frankston created the first spreadsheet—Visicalc—and licensed it to Personal Software under attractive terms. It was a runaway hit on the Apple II, motivating a name change to Visicorp and fueling growth to tens of millions of dollars as one of the world's largest software publishers. Yet Visicorp failed by 1984 and had to be sold. The main reason was a lack of focus and a failure to ensure that an excellent product in a niche market where the firm was #1 remained superior to the competition, Visicorp became distracted developing a product called VisiOn, failing to protect Visicalc's first mover advantage, which soon was surpassed in product quality and marketing energy by Lotus 1–2–3 and eventually also by Microsoft Excel.

The Canadian firm Blackberry released its first smartphone in 2000. Its excellence and differentiation were a secure encrypted network and a keyboard-centric design catering to text messaging, email, and search. The device met the needs of professionals. By fall 2010, its U.S. market share was 37% of 60 million American smartphone users. President Barack Obama felt helpless without his Blackberry. Workaholics became known as "Crackberries". Blackberry dominated the corporate marketplace—it was *best of breed*. Yet they did not protect their niche. As the tsunamis of the Apple iPhone and Android phones surged, Blackberry tried to design and build its own consumer-oriented phone with a GUI. It failed, lacking the expertise and resources needed. The company is no longer a world-class player and ceased supporting its legacy product in 2021.

Market Sizing and Market Windows

One important issue related to the choice of a market is that of size. Size is comprised of two factors—how many potential customers there are, and how much each customer will pay. Estimating the product of these quantities gives entrepreneurs a feeling for the size of the business that they could grow. Venture capitalists look for business with the potential for at least hundreds of millions (and often billions) of dollars in sales. Angels may be content with tens of millions of revenues if a company can be profitable. Family and friends will be content with much less, but they may look for the potential for profitability unless the investment is meant totally as a gift with no expectation of return.

I shall focus here on sizing the number of potential customers, which I can illustrate with several examples. I shall return to the topic of product price in Chap. 5. Here are some analyses involving our case study companies. My intent is to show how easy this can be.

Caseware International: Five minutes of web searching yields https://auditing-canada.com/Membership-Directory-&-Locator, a list of 105 auditors in Canada. A similar effort yields https://www.wpk.de/berufsregister/suche.xhtml;jsessionid=E-6390FA203092786CA034AA98AFB583A, a list of on the order of 1000 in Germany. Such research can be used to estimate the relative desirability of various international markets.

In Caseware's case, the firm had many other ways to access potential Canadian customers, as their co-founder and CEO was the country's leading visionary on the use of computers and digital technology in the accounting profession.

QLess: Amusement parks in the U.S. may be found at https://amusementparksusa.com/all-amusement-parks-usa/. 3500 American universities and colleges may be found at https://www.unipage.net/en/universities?country=224. Government and health care institutions can be researched similarly.

Chess.com: Its website gives various estimates of the number of chess players in the world ranging from 300 to 600 million at https://www.chess.com/article/view/how-many-chess-players-are-there-in-the-world. It is harder to estimate the number of serious players.

Braze Mobility: Here data took about 10 min to find, by which time I had the following: https://en.wikipedia.org/wiki/Disability_in_the_United_Kingdom, which asserts that there are about 1.2 million wheelchairs in that country, about 1 for every 56 people. This ratio could be used to estimate the number of wheelchairs in other "developed" countries; the number should be significantly reduced for estimates in "developing" countries.

Another important issue related to the choice of a market is timing. Products intended for specific markets can either be perfectly timed or too early or too late. There is a *market window* that represents perfect timing. If you are too early, customers may not be ready to adopt your solution. If you are too late, the field may by then be crowded with competitors and you may find it hard to get the attention of customers, media, and investors.

Microsoft was perfectly timed with its initial Basic product, as purchasers of microcomputer kits such as the Altair and early pre-assembled personal computers such as the Apple II needed a programming language with which to write software. They were again perfectly timed with MS/DOS, as IBM needed an operating system to run their new PCs, and also later with Windows, as the work at Xerox PARC and the announcement of the Macintosh made an operating system for PCs with a Graphical User Interface (GUI) very attractive.

Adobe also had perfect timing, as their software starting with Postscript met the needs for personal computer users and computer networks for printing. Their follow-on products were attractive to and essential for digital creators such as designers and illustrators who saw the potential of bit-mapped displays such as the Macintosh for practicing their crafts.

In many cases, being well-timed does not require precision to the extent of a few months or a year. Both Ecovative Design and Beyond Meats were founded late in the first decade of the 2010s. Concerns about the environmentally unsustainability of the planet had been widespread for quite a while, but the pace at which the world had been dealing with the problem was so slow that their timing was adequate to create vigorous, growing, and profitable businesses.

There are countless examples of being too late. By the time Visicorp realized that their distraction trying to create the VisiOn GUI operating system had allowed Lotus 1–2–3 and Excel to far surpass Visicalc, they had lost their spreadsheet dominant market position and mindshare, as well as their viability on the dominant personal computer, which by then was the IBM PC and no longer the Apple II. Blackberry's development of a mobile phone with a GUI was so slow that by the time they came out with a reliable version they were so far behind the iPhone and the many Android phones that they could never catch up.

There are also good examples of being too early. Apple made a premature attempt at a tablet, the Newton MessagePad released in 1993 and discontinued in 1998. This was a brilliant try way ahead of its time, as the iPad was only introduced in 2010. Another interesting case is the Microsoft Spot Watch, released in 2004 and abandoned in 2008, long before the Apple Watch was introduced in 2015.

Value Propositions

Having decided who the customers will be, estimating how many there are, and establishing that the market window has not closed, an entrepreneurial startup then needs to articulate a *value proposition* for the enterprise. Such statements can be qualitative. It is even more desirable that they also be quantitative. Here are examples:

Imax offers the world's most immersive theatrical motion picture experience for your customers. You can charge top dollar to attendees who will pack your theatres.

LinkedIn is the internet's dominant platform assisting almost 800 million people worldwide by career-oriented networking to help them find jobs; to maintain, grow, and enhance their contacts; and to facilitate career growth and enhance professional success.

Antibe Therapeutics will offer non-addictive medications to reduce pain to hundreds of millions of individuals worldwide.

Twitter's social networking platform assists 300 million people worldwide in communicating terse messages which share useful information and help the originators of these messages to establish and reinforce their reputations as domain experts and useful people to know.

QLess: This case is interesting because it highlights the desirability of having multiple value propositions that apply to various market niches and various stakeholders. I shall just describe a value proposition for their amusement park market.
A user of the QLess app faced with a line at a major entertainment complex such as Disneyland will typically avoid standing in two lines each day, each time freeing 45 to 90 min for wandering, exploration, and snacking. Disneyland shops will earn on a typical day $1000 more revenue because people waiting in virtual lines will drop by their booths and make impulse purchases for their kids. (Numbers invented for purpose of illustration).

Ecovative Design manufacturers shipping large quantities of boxes can reduce their environmental footprint and gain customer approval by surrounding goods with Ecovative's biodegradable packing materials.

Chess.com is the destination site for devoted students of the game of chess seeking advanced training and access to learning resources. Novice and intermediate players will find material suitable for every level of expertise, as well as continual opportunities to play well-matched opponents and to review insightful analyses of their games.

Canva is the only inexpensive tool for non-design professionals to learn and quickly be proficient in creating graphic such as web art, posters, and brochures.

Braze Mobility: Owners of wheelchairs fitted with Braze's intelligent sensors sustain wheel accidents at a rate that is 5% of that of people using standard wheelchairs (a number invented for purpose of illustration).

Product-Market Fit

Nifty technology is not enough. Extraordinary technical expertise is not enough. Working day and night is not enough. A burning societal need is not enough. Your clever technical solution must address a real opportunity or solve a real problem in such a way to bring real value to a market of customers and users. Another way of saying this is that there must be a good *product-market fit*.

In summary, I have emphasized the importance of exploiting technological opportunities to address real problems of groups of customers with significant common needs. Furthermore, you can only be successful if you have a capable management team, a talented and committed staff, a viable way of bringing your product to market, adequate capital, and luck, all of which I discuss in the following chapters. I first need to discuss one prerequisite to success, which is that your solution embody something extraordinary—an *underlying magic* or *secret sauce* that conveys *competitive advantage*.

D. Adobe Inc.

Despite being in the most exciting computer science research environment of that era, Ph.D.s John Warnock and Chuck Geschke left the Xerox PARC lab in 1982 to create Adobe. As research scientists, they had done pioneering work on computer languages for describing the printed page, which enabled communication between devices such as computers and the relatively new technology of digital laser printers. They left PARC in part because Xerox was very slow and indecisive about commercializing their Interpress technology. Warnock and Geschke saw enormous opportunity in the rapidly expanding personal computer industry, which had been turbocharged by the Apple II and the 1981 introduction of the IBM PC.

They leveraged their research experience by developing as their first product the Postscript page description language. They bolstered the product by a strategic alliance with a typesetting manufacturer to obtain the rights to common typefaces. They then developed a drawing tool to enable the design of their own fonts and released as products both Adobe Illustrator and the resulting typefaces, followed soon by Photoshop for creating images and Premiere for video editing.

Much of Adobe's growth since then has been fueled by wise acquisitions, such as Aldus with its page layout tools, and key competitor Macromedia with a suite of complementary and competitive products. We shall see this strategy repeatedly in use by successful firms. The firm has also been vigorous in expanding Postscript into the Acrobat/PDF document interchange format and in expanding and integrating its offerings into the Creative Suite and the Creative Cloud. Adobe remains today the dominant supplier of software tools to visual and design professionals as well as to many students training for careers in these professions, although they are now starting to be challenged by Canva (case S).

E. SideFX

HCR employees Kim Davidson and Greg Hermanovic did UNIX systems programming by day, but at night dreamed of applying their artistic talents and interests to work in the burgeoning field of computer animation. They left the UNIX system software firm HCR in 1985 and joined well-funded Omnibus Computer Graphics, then launched their own graphics software firm in 1987. For a name, they chose Side Effects Software Inc., enabling an evocative pun, SideFX.

Since then, they have had to compete with other firms hundreds of times their size. They have succeeded in business terms, as well as picking up five technical Academy Awards including an Oscar for contributions to the art and science of motion pictures.

Key to their success has been technical excellence, responsive attention to customer needs, and the choice of a niche market for the use of their software tools. Most computer animation is constructed (in a linear workflow) by artistic drawing, sculpting, and animating, aided by algorithmic rendering of coloured surfaces. The SideFX speciality of procedural technology allows images and motions to be created by a combination of the work of artists and the computations of machines (now also including computations by artificial intelligence machine learning algorithms). This has become more and more important as the complexity of computer-animated scenes has grown to include crowds of huge numbers of objects such as people, animals, birds, buildings, roads, trees, clouds, and mountains.

SideFX is an excellent example of a small startup that has controlled its own destiny with no venture funding and succeeded by focusing on excellence in a specific market niche. It has done so under the steady hand of a wise leader—Kim Davidson—a topic to be discussed further in Chap. 7.

F. Caseware International

Two University of Toronto engineering graduates and partners in a small accounting firm began writing a program to do personal Canadian taxes in 1981, shortly after the IBM PC was announced and after someone said to one of the partners at a U.S. trade show "You mean to tell me that you're not already doing tax returns on a computer?" The product did well, and they sold TaxPrep to a larger Quebec company in 1987.

One of the pair had a real entrepreneurial bent (he previously had started two other ventures). Thus he, as CEO, and six others formed Caseware International the following year. Another of the co-founders was a scientifically trained bookkeeping professional who became COO. The team also included highly skilled software developers. Its ambitions were great and its scope broad—to enable dramatic improvements in the processes of auditing and advanced corporate accounting.

Caseware was not an overnight success, but it grew slowly but steadily in its niche market, financing growth through founder sweat equity and retained earnings. They chose the unusual path of ignoring the U.S. market for many years, starting in Canada and then moving from one foreign market to another with capable local accounting firms as their partners, solidifying these relationships and providing capable onsite training in two or three round-the-world trips each year.

After they had achieved this significant international presence, they sometimes rebuffed suitors who sought to finance or buy them, preferring the slow and steady growth under their own control. The CEO's technical vision steered them through stages of technology growth—auditing support, desktop solutions, cloud solutions, and their current research focus, the use of AI.

In 2021, the founders exited, with an attractive sale of Caseware to a private equity firm.

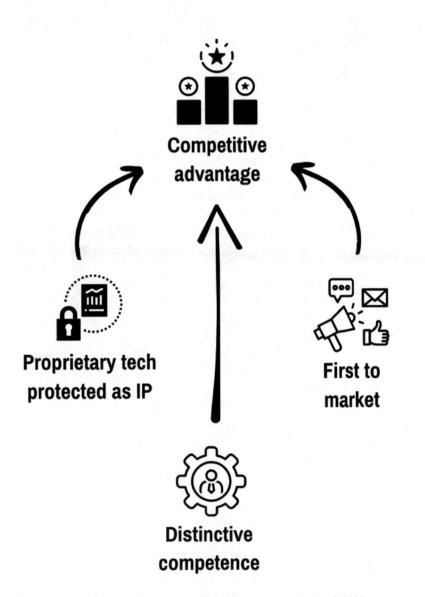

R. Baecker, *Ethical Tech Startup Guide*, Synthesis Lectures on
Professionalism and Career Advancement for Scientists and Engineers,
https://doi.org/10.1007/978-3-031-18780-3_2

Any sufficiently *advanced technology* is indistinguishable from magic.

Arthur C. Clarke (1962), *Profiles of the Future: An Inquiry into the Limits of the Possible.* Popular Library.

Ecovative Design co-founder Eben Bayer grew up on the family maple-sugar farm in Vermont. One of his chores was to gather and move wood chips from an open bunker to a gasification burner. He noticed that wet wood chips sprouted mushrooms, and their mycelium strands eventually and magically knit the wood chips together into a solid mass. In the fall term of his final year at Rensselaer Polytechnic Institute (RPI), Eben and fellow student Gavin McIntyre took a class called The Inventor's Studio, taught by Burt Swersey. Their original ideas for inventions were deemed impractical, but Eben brought to a class a solid white block of perlite particles that had been glued together with the mycelium. Swersey suggested they take the class again the next semester, and that they work on improving the ability of mushrooms to knit together waste materials to form packaging that was biodegradable. They discovered that replacing the wood with agricultural waste such as chopped-up corn stalks and husks would produce biodegradable solids imitating Styrofoam in form, function, and cost.

Microsoft and Apple have dominant market positions not just because they got an early start. All faced strong competition at every stage of growth. They have *competitive advantage* because they have something special, sometimes called *underlying magic* or *secret sauce*. They have technology and knowledge that others do not have and can therefore accomplish things that other firms cannot. This chapter discusses sources of competitive advantage for a startup such as *proprietary technology*, *distinctive competence* of its founders, and being *first to market*.

Proprietary technology may be hardware, software, genetics, or manufacturing processes. I shall discuss protecting such technology as *intellectual property* (IP). IP allows firms to solve problems others cannot, or to do things quicker or better in a way that conveys benefits to customers. I shall stress the ethical responsibility to ensure that the technology is reliable and safe for customers to use. Distinctive competence is expertise that successful firms have that their competitors lack and that enable them to provide better solutions. Some firms excel because they are the *first movers* in a new industry, or *disrupters* of an existing industry, enabling them to take a lead, build *barriers to entry* for others, and never be overtaken.

Sources of Innovative Opportunity

Proprietary technology is the result of innovation. As mentioned at the start of Chapter 1, the management consultant and author Peter Drucker has identified seven sources of innovative opportunity, which are illustrated by our case study firms.

1. *The Unexpected*

Light, small, inexpensive personal computers were unexpected. Microsoft and Apple were quick to exploit this possibility. Many firms tried to do the same, but because of other factors discussed in this book, Microsoft and Apple excelled and became dominant. As another example, nobody knew that mushrooms could knit together agricultural waste into rigid solid packing materials that were biodegradable, but the founders of Ecovative Design discovered this phenomenon, patented the technique, and built a company around the idea.

2. *Incongruities*

Restaurants have long given out wooden/electronic placeholders enabling waiting patrons to roam until a flashing light summons them to claim their tables. The technology was not scalable to handle large lines and therefore not widely adopted. QLess devised a scalable multi-industry solution that exploited the ubiquity of mobile phones. The incongruity of homeowners such as "empty nesters" who had spare space in their homes and needed additional income and more social interaction was solved by firms such as Airbnb.

3. *Process Need*

Adobe's Postscript and PDF addressed the need for seamless interchange of pages and documents between computer hardware, software, and printers. Twitter enabled sending out via the internet very short text messages—status messages or bites of information—as an alternative to social media posts or podcasts. Wikipedia solved the need for ever-expanding encapsulations of human knowledge which was validated and fact-checked by motivating thoughtful people worldwide to do this task without the need for paying them.

4. *Industry and Market Structures*

Beyond Meat took advantage of increasing societal concern about the ethics and processes of producing and consuming meat to create meat substitutes that fit the existing distribution and retailing of meat products perfectly. Nuula entered the crowded market of software solutions for small businesses on the belief that what was lacking and needed were timely infusions of both critical information specifically relevant to their business situation and small amounts of working capital.

5. *Demographics*

MasterClass anticipated that the growing hunger of educated people for truly extraordinary adult education would fuel a desire for learning experiences to be taught by the very few individuals in the world authentically considered to be masters in their discipline or craft. Winterlight realized that a novel ability to diagnose and monitor cognitive decline through the analysis of speech would find growing markets as the percentage of older adults continued to increase and cognitive decline became more widespread.

6. *Changes in Perception*

By 2003, it was clear that more and more activities were moving online, hence LinkedIn created a web platform for professionals to support career advancement and networking. Chess.com decided in 2007 that advances in algorithms for playing and analyzing the game could be applied to support regular online play and learning by fans of the game.

7. *New Knowledge*

The founders of Blue Rock Therapeutics recognized that stem cell technologies were enabling the genetic engineering of cells with revolutionary medical potential. The founder of Braze Mobility became a world expert in applying computer vision to improve wheelchair mobility and safety, hence it was logical for her to take this expertise and bring products to market.

In many cases, successful or promising startups leveraged several sources of innovative opportunity. For example, Braze like Winterlight takes advantage of demographics. Ecovative Design like Beyond Meat exploits the increased awareness of environmental dangers. LinkedIn and Twitter's rapid growth is enabled by the universality and appeal of social media, as well as by the characteristics of the social media marketplace.

Reliability and Safety

These examples show how tech firms often derive their competitive advantage from some proprietary underlying magic. Yet ethical "tech startups" must be honest with themselves and with customers if the technology really works. Does it achieve the stated goals? Is the product line consistent with corporate values and its vision and mission? Is the technology reliable? Does it deliver consistent results? Is it safe?

A classic case of an organization prioritizing markets over safety is that of the computerized radiation therapy machine Therac-25 from Atomic Energy of Canada Limited. Six incidents of massive overdoses of radiation occurred between 1985 and 1987, typically accompanied by patient complaints that they were burning, complaints that were usually ignored. Four cases ultimately resulted in patient death; the other two patients suffered serious injuries. It took two years for regulatory agencies in the U.S. and Canada to halt use of the machine.

A recent example of a tech company, albeit far from the startup stage, behaving unethically in choosing profits over customer safety is Boeing with respect to its 737

Max airplane. Partly in response to a perceived competitive threat from the EU's Airbus, Boeing replaced the engines with new ones that were more fuel-efficient. To do so, they had to change the positions of the engines, with the result that it made the plane aerodynamically unstable. Nonetheless, Boeing rushed the new product to market in 2017 without including adequate technical safeguards. They also hid the need for more pilot training. The result was two crashes within minutes of takeoff that cost hundreds of lives. Only in 2019 was the plane finally grounded so that major repairs could be made. This case also illustrates a situation in which the current management had values far different from those of the founders and from the previous team.

Proprietary Technology and Intellectual Property

A key word to most innovation capable of commercialization is *proprietary*. The Cambridge Dictionary defines this is "relating to owning something". In each case, our companies possess technology that their competitors do not have, although competitors likely have something similar. Our firms typically succeed because their proprietary technology is much better. I emphasize that it must be **much better**. Venture capitalists sometimes describe this goal as a 10X (tenfold) advantage. Even if proprietary, me-too products that are very similar to others will typically fail, unless there are other significant sources of competitive edge, such as company size, geographical proximity, superior customer service, compatibility with other systems, or price.

Proprietary technology is *intellectual property* (IP). Because it conveys competitive advantage, it must be protected so that others cannot steal it or copy it. There is an alternative approach not involving proprietary information called *open source* and *open access*. Although I am not discussing this strategy in this short book, I provide some references in the Further Reading appendix. Some people feel that intellectual property is selfish and unethical. I understand this point of view, but do not adhere to it.

It is important to stress that the goal of entrepreneurship is to create **value** and not just technology, so entrepreneurs wishing to compete in some marketplace can do so by acquiring technology developed elsewhere, turning it into a product, and then marketing it. A good example is Beyond Meat's licensing of a patent for "plant-based meat" (developed by university professors).

Trade Secrets, Copyright, Patents, Trademarks

I shall discuss four methods of protecting intellectual property. *Trade secrets*—your knowledge that is valuable and that others do not possess—may be kept secret if you make reasonable efforts to protect them. *Copyright* guards written expressions of ideas such as writing and computer programs from being used by others without your permission. *Patents* convey similar protection to the processes and methods you have invented;

some patents are licensed, yielding income for the inventor and enabling useful techniques to be applied more broadly. *Trademarks* convey protection to the words and images you use to describe your inventions, creations, and products.

Trade secrets: Examples of trade secrets are the formula for making Coca Cola, and the blend of spices used in Kentucky Fried Chicken, Tech startups have trade secrets, such as the algorithms used in their software, their business and marketing plans, and their customer lists.

For a trade secret to be protectable by law, it must have value. The knowledge that you will ship a bug fix tomorrow has little value, but the schedule for introducing new products over the next year could be important to a competitor. Your method of storing and retrieving customer data likely has little value to a competitor, but both the architecture of your software and the names in your customer list are valuable.

Trade secrets cost money and ongoing vigilance to protect. For courts to agree that a trade secret has been stolen, you must have made reasonable efforts to protect it. Good examples are having locked doors to your building and having security guards and an after-hours sign-in procedure in large buildings; using passwords and cryptography for securing your computers; and making employees and contractors sign agreements ceding ownership of their work to your firm and agreeing that they will not disclose secrets or give code to others after they cease working for you. One of our case study startups, Beyond Meat, has been involved in a bitter 5-year legal battle involving claims of misappropriation of trade secrets with a former manufacturer of its products.

There are limits to methods of protection. For example, onerous non-compete clauses that would create unreasonable restrictions on future employment are often deemed unreasonable by courts.

Copyright: Copyright gives an author the exclusive right to reproduce, display, and distribute an original work that is fixed in a tangible medium of expression, i.e., it cannot just be an idea. Examples are essays, books, musical scores, drawings, icons, sculptures, and computer programs. You can copyright the code expressing an algorithm but not the algorithm. Copyright will protect someone from making and using an exact copy of your creation, that is, from copying the code for your social media app, but not from creating their own similar social media app.

Copyright costs nothing to declare. Copyright © vests when you create the symbol and fix it to the medium, as for example by writing it on every page of your code. You don't have to file anything. I can copyright a book or a program or a diagram by writing or typing "Copyright © 2022 by Ronald M. Baecker" on the artifact. Better protection in terms of potential damages is afforded by registering the copyright and using the symbol ®, but this is rarely done for software. Copyright affords protection for a long time, on the order of a century.

Copyright protects writing; patents protect processes. Software is both writing and process, so both methods apply.

Patents: Patents are for inventors, and grant them the exclusive right to make, use, and sell their inventions. This means that they can prevent others from using the inventions or can license others to use them on mutually agreeable terms. Alternatively, they can sue firms that are using or selling technology that is infringing their patents and seek an injunction to prevent them from doing so without paying a license fee.

Innovative products that are manufactured in traditional ways are usually protected by patents. For example, Antibe Therapeutics has patents that cover every possible element that can be protected. Beyond Meat and Blue Rock Therapeutics each have one key patent. Ecovative Design has obtained almost 40 patents; Imax has obtained almost 100 patents.

Patents play a key role in biotech. The FDA regulates drugs using patents, extending the life of patent protection to enable drug developers to gain additional exclusivity because many drug programs do not deliver working drugs, and development and approval cycles are substantially longer than for other intellectual property-driven industries.

Mathematical formulas and equations are not patentable, so programs were not thought to be patentable at the onset of the computer industry. This changed in the 60s. Now, 63% of patents issued by the U.S. Patent Office are for software.

A good example is the 2000 lawsuit brought against Blackberry by NTP, Inc., a Virginia-based holding company (a *patent troll*, a firm that buys up IP from bankrupt firms and looks for companies to sue for patent infringement) specializing in areas including wireless email. NTP claimed that Blackberry's wireless email system infringed their patents, and asked for the payment of damages and for the cessation of sales of Blackberry devices in the U.S. Given Blackberry's strong market position at the time, the U.S. Justice Department filed a brief arguing that Blackberry service needed to continue because it was essential to national security. The case was finally settled in 2006 with Blackberry paying NTP US$612.5 million.

Another example is the 2008 lawsuit against Apple by Mirror Worlds, LLC, a subsidiary of a patent troll that owned the IP of eminent Yale Computer Science Professor David Gelernter, who is also known for having a hand blown off by the Unabomber. The suit alleged that certain user interface aspects of the Macintosh and iPhone operating systems infringed patents awarded to Gelernter based on his university research on time-ordered file systems which he had unsuccessfully tried to commercialize early in the century. I was a fact witness for Apple in the trial, as Gelernter's work was somewhat similar to a research project I had undertaken at the University of Toronto. At first, Apple was ordered to pay US$625.5 million in damages, but the decision was reversed on appeal. The patent troll sold the IP to another patent troll, that was successful in 2016 in getting a US$25 million settlement from Apple as well as amounts from other firms including Microsoft.

Processes or machinery patented must be man-made, useful and enabling, providing users practical advantage. They must also be novel and non-obvious to "someone skilled in the art". Patents are not easy to obtain. Patent applications are lengthy documents with text and many diagrams explaining the invention and long lists of claims about what it enables. Most entrepreneurs will need to hire patent attorneys to do the work, but will still have to invest significant time working with the lawyers.

Entrepreneurs who come from academia face an additional hurdle, as they will often want to publish their ideas and inventions. In the U.S., patent applications are granted to the first person to file, so publication before at least filing a provisional application (see below) endangers your ability to receive patent protection.

U.S. patents are typically valid for 20 years, although there are vast differences world-wide. Costs can easily exceed US$25,000 for filing in one country; each country must be dealt with separately, although there is an expedited process for countries in the European Union. The process is slow and can take upwards of 5 years before the patent is granted. Provisional patents cost 1/10 of this and give the inventor one year to file the complete application. Because of the work, time, and cost involved, startups with limited funds will usually forgo the opportunity, although being able to say Patent Pending confers credibility for sales purposes and for attracting venture capital.

Trademarks: These are names or symbols used by companies to identify, distinguish, and indicate the source of goods. They can make use of a variety of sensory modalities. Good examples are the word "Word" (used by Microsoft to identify its word processing software), a design such as Exxon's tiger, a sound such as NBC's chimes or the music played in broadcasts of U.S. professional football games, a slogan such as Avis's "We try harder", a sculpture such as the Mercedes Benz hood ornament, or a colour such as Fiberglass Pink.

There are various classes of trademarks—descriptive marks such as Frosted Flakes, suggestive marks such as Coppertone, arbitrary marks such as Apple, and fanciful marks such as Sony. Marks used in a generic way lose their protection, such as has been the case with "kleenex", "escalator", and "xerox", and will likely soon be the case with "zoom".

Trademarks are inexpensive to register—$400 for a single mark in a single class, and protection lasts 10 years. They must be policed, asking infringers to cease or face a lawsuit, or the trademark is lost.

Tech firms use trademarks both for their company names and for the names of products, with the goal of ensuring that competitive and inferior products are not purchased because of buyer confusion, and with the goal of building a brand (Chap. 4).

Beyond Meat has filed for over 100 trademarks including Beyond Chicken, Beyond Burger, Beyond Beef, Beyond Sausage, Beyond Tuna, Beyond Crab, and Beyond Deli. The goal is to ensure a dominant and consistent brand and mindshare with customers and potential customers.

Distinctive Competence

IP by itself is insufficient. Speed and agility are vital—neither is proprietary nor techno-logical. It is essential to be first or early to market; have momentum, strategy, and tactics to gain and maintain competitive advantage; introduce innovative products and features rapidly; deliver a superior user experience; support products effectively and humanely; and understand your market and customers much better than your competitors do.

To achieve this, startups need *distinctive competence*—knowledge which they pos-sess, and their competitors lack or have only to a lesser degree. Let's see how this mani-fested itself in our case study firms.

Adobe founders Drs. Charles Geshke and John Warnock had spent several years at Xerox PARC developing and refining page description languages, so were able to take their knowledge and wisdom and quickly develop Postscript as a superior product.

Caseware founders had an excellent understanding of the needs of accountants and auditors. The CEO's reputation as the Canadian accounting technology visionary helped Caseware build a brand as an innovator and market leader.

Founder of Antibe Therapeutics Dr. John L. Wallace has been exploring the role of hydrogen sulfide in gastrointestinal health and as an anti-inflammatory and pain killer for over 20 years.

Co-founder of Blue Rick Therapeutics Gordon M. Keller has been researching stem cells and how to use them in the treatment of disease for almost 30 years.

By the time Dr. Pooja Viswanathan founded Braze Mobility in 2006, she had spent 10 years doing graduate and postdoctoral research on intelligent wheelchairs that could help elderly users avoid collisions and do better wayfinding, thus providing her with superior technical skills as well as a good understanding of potential customers and their needs.

Nuula's founder Mark Ruddock brought to his 2021 startup over three decades of entrepreneurial and CEO experience and two successful exits from successful informa-tion technology growth firms focusing on financial services.

There is another kind of competence very different from deep expertise in a startup's domain. I shall call it a *nose for disruption opportunities*. This is the ability to look at an area, often as an outsider, and imagine ways of doing things nobody else ever envi-sioned. In some cases, your collaborators will think you are hallucinating, but you have the courage, the imagination, and the arrogance to proceed. Here are some examples.

Steve Jobs of Apple had an extraordinary design sense and feel for the market, both in terms of long-existing products such as computers, and with respect to new technologies such as mobile phones and tablets. He was able to imagine new approaches to technol-ogy, such as low-cost mass-market personal computers with GUIs, and ways to do busi-ness, such as selling individual songs over the internet and establishing stylish stores to sell only his computers.

Airbnb's founders imagined an entirely new method of booking and renting rooms. Elon Musk of Tesla, although originally not an expert on automobiles or batteries, immersed himself in these technologies and in the processes of manufacturing and made a huge and in retrospect correct bet that the time was right for electronic vehicles.

Uber's Travis Kalanick totally changed the experience of finding a ride in cities, yet his actions with respect to competitors, regulators, customers, employees, and drivers is a classic case of unethical entrepreneurship. Uber spied on competitors and used dirty tricks such as calling its main competitor Lyft for imaginary pickups to damage Lyft's reputation and good will with drivers. Cities such as San Francisco and New York (where six taxi and limo owners committed suicide within one year) tried to regulate Uber, but it opposed regulation vigorously and ignored legal orders. Uber raised prices exorbitantly in respond to high demand and seemed to be insufficiently concerned for the safety of its passengers. Corporate culture has been widely characterized as one of "toxic masculinity", bullying, sexual harassment, and misogyny. Kalanick's inhumanity also manifested itself in Uber's long refusal to allow drivers to be tipped. He finally resigned in 2017 after 7 years as CEO.

In all these cases, superior experience, knowledge, and intuition has been accompanied by drive, hard work, and a passion to be the best.

First Mover Advantage, Scale, and Barriers to Entry

Also, these cases illustrate the importance of being early in the market. One typically speaks of *first mover advantage*, although it is not essential to be **the first**. What is essential, according to LinkedIn founding CEO and successful venture capitalist Reid Hoffman, is to be the first entrant to achieve significant market penetration, or *scale*.

There are several aspects to a first mover advantage. First and early entrants into a marketplace have lead time to establish their brand before they have significant competition. They have first access to valuable knowledge of customer needs and market characteristics. They can acquire reference accounts before others do the same.

The personal computer revolution started with the 1975 announcement of the Altair microcomputer kit and the founding in the Bay area of the Homebrew Computer Club. In April 1977, San Francisco hosted the first West Coast Computer Faire. Although there was still some interest in kits, among the exhibitors were a few companies with stand-alone home computers, most notably Commodore with its PET, Radio Shack with its TRS-80, and Apple's Steve Jobs and Steve Wozniak with their Apple II. Within a few years, only Apple remained viable, as it was able to exploit its fast start, superior graphics, and availability of software such as Visicalc to achieve temporary dominance, although that went away with the announcement of the IBM PC in 1981.

Visicorp also had a first mover advantage with the spreadsheet. Initially, their brand **was** the spreadsheet—a new product category. Yet their ambitions led them into a development swamp called VisiOn, which distracted them. They did not protect their flank and were assailed in two directions. Microsoft and a startup called Lotus Development focused on the rapidly expanding market for the IBM PC and PC-compatible personal computers and also developed better products. Microsoft developed Excel, with superior functionality and user interface, the latter coming via the expertise of developers who had worked at Xerox PARC. Lotus created 1–2–3, establishing a new product category, a 3-in-1 productivity solution, incorporating a spreadsheet, a database manager, and a tool for turning financial data into charts and graphs. Visicorp thus sunk rapidly into oblivion.

LinkedIn, Twitter, Airbnb, and MasterClass all were pioneers in their product category, then establishing dominant positions in their marketplaces as *best of breed*. Antibe Therapeutics, Ecovative Design, Beyond Meat, and Blue Rock Mobility were early entrants in their marketplaces. Although there had been prior attempts at more immersive motion picture theatrical experiences, Imax was unique, and established its own product category.

Firms that successfully exploit early entrance into a marketplace and that achieve dominant positions then benefit from *barriers to entry* that obstruct the entrance of potential competition. The early entrants have superior knowledge of markets and customers. They have long lists of satisfied reference accounts and impressive market share. They provide a satisfying professional climate for software developers wanting to contribute to an industry's leading products. They also find it easier to acquire "shelf space" on websites to get their message across, and to get reviews from journalists who want to describe and evaluate only the leading products in a product category.

In summary, protected intellectual property, distinctive competence, and speed to market all convey competitive advantages. Yet these advantages may be illusory if assumptions about market size are exaggerated, if predictions of customer needs are unrealistic, and if assumptions about product quality are hallucinations. Hence the need for validation, which is the topic of the next chapter.

G. New York Times Reinvented

The newspaper industry has been almost destroyed by internet communications and the availability of news on the world wide web. From 2005 to 2021, for example, 2200 American local papers closed due to significant reductions in traditional ad revenue. From 2008 to 2020, over half of American journalists left the workforce.

Newspapers began a technological transformation of production processes in the 1960s with a transition to digital typesetting, and in the 1970s with the use of online composition and editing of stories. News delivery via the web and news reading electronically was an even more dramatic change starting in the 90s.

The New York Times, founded in 1851, faced the same challenges, and began its transformation to an agency targeting both digital media and print with the NYTimes.com website in 1996. At first the digital content was free, and much still is, but the Times erected a pay wall in 2011. Echoing a theme of Chap. 3, this was driven by a relentless focus on the customer. Growth of the digital business has therefore been driven by the creation of great content including much that is non-traditional, such as podcasts and interactive stories, as well as a renewed emphasis on creativity in support of activities such as cooking and playing word games (see Case Z: Wordle), which now is responsible for over 1 million subscribers.

Much of the growth has been architected by CEO Meredith Kopit Levien, only the second women to head a business historically run by old men, and the youngest in history, echoing the theme of diversity I will discuss in Chap. 7. Results are outstanding. The Times currently has 800,000 print subscribers and over 5 million digital-only subscribers, well on the way towards its very ambitious goal of convincing 10 million customers to pay for online quality journalism.

H. Shaw Industries Reinvented

Shaw Industries Group, Inc. supplies carpet, resilient, hardwood, laminate, tile, and stone flooring products and synthetic turf for homes and businesses. Founded in 1967 and headquartered in Dalton, Georgia, it became a wholly owned subsidiary of Warren Buffett's Berkshire Hathaway, Inc., in 2000. It has grown via innovation and acquisitions steadily over the past 55 years.

In the late 1990s, it began to reinvent itself with the goal of designing products that support a circular economy, where materials and products can either be used for as long as possible or refurbished or recycled. Sustainability became a goal and one of the dominant values animating the firm. More specifically, it launched in 1999 the EcoWorx® carpet tile, the industry's first fully recyclable, PVC-free carpet tile. It became one of the first Cradle to Cradle Certified® products of any type in the world and the first of any flooring product to be certified.

Shaw has engaged in Cradle to Cradledesign and product certification for over 20 years. Today, almost 90% of its products are certified; it recently had one of the first products in the world to be certified to the new, most rigorous standard version 4.0. Shaw has also created manufacturing facilities using as little energy as possible.

According to Shaw's Director of Sustainability Communications: *"As neither design nor performance of EcoWorx was significantly different than what we already offered, we infer that it was in fact the sustainability attributes of being made with a focus on green chemistry and being fully recyclable that led to the dramatic growth in sales. Before we launched EcoWorx, Shaw was the #4 or #5 carpet tile manufacturer; EcoWorx has led us to being the #1 or #2 carpet tile manufacturer."*

I. Desire2Learn (D2L)

John Baker founded D2L in 1999 while a third-year engineering student at the University of Waterloo with the vision of transforming teaching and learning. After early success with Canadian adopters such as the University of Guelph, D2L landed its first sale of its learning management system (LMS) with the state of Wisconsin in 2003. Success in landing the account was enabled by several all-nighters of the dedicated and talented technical staff in response to questions and challenges posed by Wisconsin employees.

D2L has faced continued competition and disruptions from world events as it strove to become and stay an industry leader. Originally, Blackboard was the market leader, ten times bigger than D2L. After receiving a patent in 2006, Blackboard sued D2L, consuming millions of dollars and huge management attention until a federal appeals court voided part of the Blackboard patent and the two firms settled out of court. Since then, D2L has amassed a portfolio of 70 patents to lead the LMS marketplace in IP protection and defense.

The LMS marketplace has long been a battleground of two different paradigms of modern software, proprietary systems and code that is open source. Although Wisconsin announced in 2017 that they were adopting the Instructure Canvas open source LMS, D2L is battling its open source competitors with vigorous development of cloud solutions.

A more recent challenge arose with the COVID-19 pandemic in 2020. At first a challenge due to increased use and greater support costs, and a reduction in universities seeking new technology, the pandemic has now resulted in more institutions seeking improved solutions.

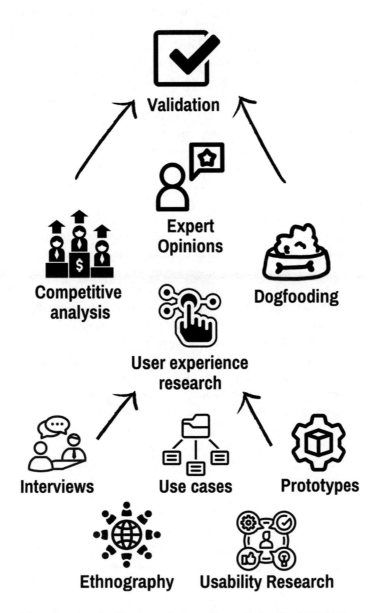

R. Baecker, *Ethical Tech Startup Guide*, Synthesis Lectures on
Professionalism and Career Advancement for Scientists and Engineers,
https://doi.org/10.1007/978-3-031-18780-3_3

Don't make me think.

Steve Krug (2013), *Don't Make Me Think, Revisited:*
A Common Sense Approach to Web Usability.

Braze Mobility founder Pooja Viswanathan's doctoral and post-doctoral research
focused on smart wheelchairs. She engaged stakeholder extensively, attending and host-
ing accessibility events and focus groups, and networking with researchers, clinicians,
tech distributors, and potential users. She learned much, most notably that users did not
want the wheelchairs to take total control, and that a smart wheelchair's best mode of
operation would depend heavily on both the users' physical environment (home or insti-
tution) and their circle of care. In founding the firm in 2016, she initiated a detailed
planning process, integrating the results of research studies with information gleaned
from customer interviews to pivot the company concept from the originally envisioned
highly automated wheelchair to "spot sensors" that could be attached to existing
chairs—a simpler, more sensible, and more tractable approach, and one that allowed the
company to face fewer regulatory hurdles.

We all encounter daily technology that does not work as intended, that is frustrating
and forces you to think hard, that you cannot use effortlessly. Even the most brilliant
innovations, turned into new ventures by great entrepreneurs, have no guarantees of success. Years of time and millions of dollars are often required to develop products, ensure
they meet user needs and are safe and usable, know if they will find markets, and see if
they will anchor profitable businesses. Since we cannot predict the future, we need ways
to test ideas and plans to gain some insight into whether our product and market expectations are credible. This chapter presents methods to validate the viability of planned
products, and to ensure that people can use them naturally and effortlessly.

I shall discuss methods that focus on the competitive landscape (*competitive analysis*)
and on the prospective user. Latter methods include *market research*, customer research
with *ethnographic* field methods, *prototyping*, *user experience research*, *quality assur-*
ance and *dogfooding,* and a constant focus on achieving *customer success*. I shall stress
the need to ship versions of your product early and often, even when still immature, to
get lots of customer feedback.

Competitive Analysis

When I entered the software industry in 1976, competitive analysis was difficult.
Software packages cost thousands of dollars, so you could not reasonably buy one just
to "kick the tires". Hardware cost tens of thousands of dollars, so it was difficult to try
software X on hardware Y. Product evaluations were published in magazines, but the
most thorough analyses were published in market analysis documents costing hundreds
or thousands of dollars.

It is much easier today. Most apps cost tens or at most hundreds of dollars. Almost all have free demo versions that you can try for a week or a month. In many cases, demo versions have reduced functionality, but you can see enough and kick the tires sufficiently to get a good feeling for the product. Expert analyses are still expensive, but many evaluations appear in trade magazines and on websites. Social media report numerous instances of what customers think and what their experiences have been.

For products that are not software, analyses must be assembled based on your firm's ability to try the products yourselves as well as what you can learn from third-party reviews, customers, or industry experts who you can try to meet at industry trade shows and access via email or phone.

A competitive analysis examines three kinds of issues. The first is a catalog of *functionality*—what the product does. The second is a description of the *user experience* (UX)—how satisfying or frustrating it is to use the product. There is also the *customer experience (CX)*, focusing on the firm or individual who buys the product—often not the user. CX includes strengths and weaknesses of the firm offering the product. Attributes for public companies may include measure of financial performance such as net worth, profitability, revenue growth, and market share in specific domains.

A competitive analysis chart comparing your product to two others would like this this:

Feature	Competitive product 1	Competitive product 2	Your product
Functionality feature 1	Rating	Rating	Rating
Functionality feature 2	Rating	Rating	Rating
...			
Functionality feature i	Rating	Rating	Rating
UX feature 1	Rating	Rating	Rating
UX feature 2	Rating	Rating	Rating
...			
UX feature j	Rating	Rating	Rating
CX feature 1	Rating	Rating	Rating
CX feature 2	Rating	Rating	Rating
...			
CX feature k	Rating	Rating	Rating
Overall	Rating	Rating	Rating

This table need not be huge, as it is most effective when it fits on one page. Functionality measures typically occupy half to two-thirds of the rows, with the remainder split between UX and CX. Ratings in the competitive matrix may be binary, i.e., whether a feature is present or not; qualitative, e.g., excellent, good, fair, poor, dreadful; or quantitative, e.g., a quality indicator on a scale of 0 to 10 or a 4-level scale of "+++",

"++", "+", or "" (there is no entry in that field of the competitive matrix. Choice of the rating method depends upon the category rated and the quality and quantity of information supporting the rating.

An example of a competitive analysis for this book appears in the sample pitch deck for this book shown in the first appendix.

Here are considerations for competitive analyses of two of our case study firms:

Beyond Meat: The competitive position of one of their products must be compared and judged with respect to other sources of veggie-based imitation meat or with respect to traditional meat products.

Functionality features would include nutritional qualities of the product and the absence of harmful additives. UX attributes would include taste, "juiciness", odor if any, and ease or other attributes of cooking. CX attributes include price; suitability for use in different countries, cultures, and languages; breadth of distribution and availability worldwide; and the credentials and credibility of the company.

MasterClass: Consider the MasterClass products that teach aspects of cooking. Their offerings as of January 2022 number 19 cooking classes, including 7 that are generic, and 12 specialized programs on topic such as Indian cooking and on Texas-style BBQ.

A small amount of web searching yields competitors including dedicated online cooking schools such as America's Test Kitchen, media organizations such as the New York Times and the BBC, "universities" such as Top Chef University, Massive Open Online Course (MOOC) providers such as Udemy, and gig economy providers such as Airbnb. A challenge for a MasterClass competitive analysis is to decide which of the 50–100 competitors are strong and viable.

Functionality features would include such attributes as breadth of content, depth of content, length of lessons, and expertise of the instructors. UX attributes would include clarity of presentation, quality of the videos, empathy of the instructor, and ability for students to interact with the instructor. CX attributes would include price; ability to receive a certificate if one is taking the class for professional reasons; and longevity, solidity, and prestige of the offering institution.

There will be wide variety in the organization and presentation of competitive analyses. What is essential that the contents be valid, based on reliable intelligence; that they be truthful, not omitting salient information; and that they be relevant and comprehensive enough for a company's senior management or for product managers to assess whether their firms can possibly compete successfully. Positive decisions are of course predicated on the assumptions that development produces competitive products and keeps them in good shape and that marketing has the skill and resources to build and maintain customer awareness.

How much do you need to know about the competition? Everything! Since that is impossible, the practical answer is that you should gather as much intelligence as possible within the constraints of your time and budget. Also essential is that you revisit the data from time to time, as your serious competitors will be innovating and improving both their product and their market position, and you need to be aware of their successes, failures, strengths, and weaknesses.

Differentiation and Positioning

Insights you can get from a thorough competitive analysis are an understanding of the strength of your competitors, as well as details of their respective strengths and weaknesses in various areas. This information can help you achieve *differentiation* from others offering similar products. Here are some examples.

Imax: Because of its superior sense of immersion, Imax has very strong differentiation from other forms of theatrical entertainment and their venues, such as motion pictures shown in conventional movie theatres and concerts performed in concert halls or stadiums.

NYTimes.com: The digital version of the New York Times stands out from other digital news sites because of its long status as one of the world's most comprehensive, reliable, and thoughtful sources of news, and because of the cleverness and agility with which it has assembled compelling digital media for its site.

LinkedIn: Its primary differentiation from other social media sites is its clientele—professionals—and its purpose, which is networking and career advancement. This leads it to differing in many details. For example, it is heavily oriented to text, and to communications longer than are typically found on social media such as Facebook. It also provides privacy-sensitive tools for connecting with other professionals with shared occupations or interests.

Antibes Therapeutics: Its products are still in clinical trials, but it will have to differentiate itself against other drugs for alleviating pain by demonstrating stronger effectiveness and/or fewer harmful side effects.

Chess.com: Its primary differentiation from other online chess sites such as Chess24 and Chessworld.net is the quantity and quality of teaching materials, including insightful game analyses and both text and video analyses. It also provides smooth support for playing against either bots or humans, who may be specific friends or random players at about the same level who may be located anywhere in the world.

MasterClass.com: Its primary differentiation is its selectively, featuring classes from individuals who are demonstrably the top in their field. Its emphasis on superlative production values in the instructional videos is consistent with its high-end brand.

MasterClass.com also differentiates itself with its charging model, which is not by the instructional unit of the class, but rather a monthly subscription cost of CDN $20 per month for all the learning customers want to do.

Nuula differentiates itself by the relevance and utility and timeliness of the reports to its small business customers, and by the rapid infusions of small amounts of capital for emergency use by these businesses.

Why is differentiation so important? Except in the first few months of a product that invents a new category and hence has no direct competitors, there will always be a crowded competitive landscape. You must differentiate by *positioning* yourself in the landscape of all the competitors. You need this so that potential customers can pick you from the crowd and know why they should buy your product. For example, they can identify LinkedIn as the only social medium catering to professional advancement, Chess.com as the best online chess site for improving your game, and Masterclass as the online instructional website with the instructors who are at the very top of their fields.

Finally, the differentiation should tell you what you need to do to maintain your position. Visicorp should have but failed to maintain its position as the vendor of the world's best spreadsheet. Blackberry should have but failed to maintain its position as the mobile phone for professionals. On the other hand, Adobe has continued to be the source of tools for visual creators such as designers and filmmakers. Wikipedia has strengthened its position as the world's most comprehensive and accessible encyclopedia. SideFX has continued to provide superior tools for teams of artistic and technical computer animators. LinkedIn and Twitter have continued to be the social media for professionals and for the widespread dissemination of short bursts of information, respectively.

User Experience Research and Tech Usability

Think of starting a new venture as preparing to run a race. You only want to enter the race if you have some chance of success. Success to you may be defined in various ways—finishing first, finishing in the top 10, finishing in the top half of participants, finishing at all (in the case of a marathon), doing better than your running buddy, or doing better than you did the last time.

Competitive analysis in this example is the gathering of intelligence about others planning to enter the race. For success criteria dealing with winning or doing well enough relative to the field, you need to understand how strong the other runners will be. Do you have a chance to win or place in the top 20%?

Yet you also need to know and understand the characteristics of the race and the track over which the race will be run. How far will you need to run? How hot will it be? How high will the humidity be? What will be the altitude? What is the topography, in other words, will there be hills to ascend and how steep will they be? If your goal is place in the top 10, the answers to these questions will help you decide if you have a chance and how you should train so that you have a realistic possibility to succeed.

Tech startups must also understand the landscape of their race to the top. Are there sufficient potential customers? What are their characteristics? How do they currently do the job (working or learning or playing) that you hope to assist or replace with your technology? Do they need your product, and, if so, to do what tasks? How important are these tasks, and will improving their performance on the tasks save them money or save lives? What are they likely to think of your technology? Do they have ideas for improvement that could help you build a better product? Answering these questions about prospective as well as actual customers is the job of *user experience research.*

Ethical tech startups will prioritize the user experience of their customers. They can base their designs on deep research eliciting an understanding of how potential users work, learn, and play, and how their product could meet user needs and solve user problems. They can consider during the design process all the ultimate stakeholders of a technology, and not just the financial returns to a firm's shareholders. They can try to ensure that the goal of benefiting the major stakeholder does not do damage to others. Their user experience research can help them understand where people experience difficulties with their products and help them improve their designs so that the problems do not recur, so that their products *usable*, i.e., easy to learn and easy to use.

Ensuring that a product makes user needs and is usable gives it a chance for success in entrepreneurship races and marathons.

Stakeholders and Personas

Despite the word "user", good user experience research investigates more broadly than just the actual user of your product. It considers all the *stakeholders* who will influence and be influenced by the adoption of your product.

I shall start with **Adobe**, focusing on its Premiere digital video editing software, first introduced in 1991. This is a tool for film editors to help them in digitally splicing together film clips, trimming them of unneeded material, and modifying and overlaying them with special effects such as fades, dissolves, and colorization. Film editors work in an ecosystem of multiple stakeholders such as producers, directors, camera people, actors, agents, and distributors.

In the case of **Airbnb**, there are two users of the app, the individual or family desiring accommodation, such as a traveler coming to a city for tourism or work, and an individual offering to provide accommodation, such as a person making a room or apartment available in his or her home. There are many other stakeholders who are affected by the act of using Airbnb. Most important are a homeowner's neighbors, as many have been disturbed and inconvenienced by a regular flow of strangers into a neighborhood or building and by the loud parties that some Airbnb renters have thrown. Many apartment complexes have therefore erected barriers against Airbnb use, as have some cities, who have also been upset by reductions in tax revenues they earn from hotels. Hotel chains have noticed a reduction in revenues and have reacted to this competitive threat by starting their own businesses offering accommodations in a manner like what is offered by Airbnb.

Braze Mobility: The user of a Braze Mobility enhanced wheelchair is the owner of the wheelchair, the individual whose mobility is enhanced by sensor warnings about obstacles. Yet there are others whose lives may be changed by such use. One example is other seniors who are in wheelchairs or on foot, who may be threatened by what they perceive as new careless speed by Braze users who now feel safer. Another stakeholder is the management of a senior care facility or nursing home, who must consider the safety and comfort of all the residents. Since many users of wheelchairs work with clinicians and therapists whose job is the enhancement of wheelchair users' welfare, these individuals also are stakeholders. Other stakeholders are family members, insurance companies and governments that may pay for the technology used by people with mobility challenges, as well as the companies manufacturing and selling wheelchair technology.

Drisit: Drisit tourists will take temporary ownership of a drone that flies throughout an environment under the control of a user. Consider a Drisit tour of New York City. Airports are of course concerned that a drone may hit an airplane. Pedestrians enjoying Central Park may have their tranquil experience disrupted by drones. Residents of apartment buildings may have privacy concerns. Parents may worry for the safety of their children. The city will need to deal with all these concerns.

Hence product and market planning must think broadly about the full spectrum of people and institutions involved in and affected by new technology use to anticipate the personal, societal, legal, and financial implications of deploying innovative products in homes, workplaces, schools, and urban and rural environments.

All stakeholders need to be considered before launching a new product or venture, but the *user* or *users* need special consideration. We do so by constructing *personas*, or idealized descriptions of them. Are they primarily of one gender or another? How old are they? Do they live with partners and have children? In what kind of neighborhood do they live? How much education have they received? What jobs do they do? What level of responsibility do they have in those jobs? Do they have special interests? Do they have certain challenges or disabilities? What passions motivate them? Are they risk takers?

If your customers are corporations, there is a corresponding set of questions. What businesses are they in? How large are they? Where are they located? Who are their customers? How do they attract and retain their customers? Are they growing? Are they profitable? What kinds of technology do they need and use?

Adobe: Users of their Premiere product are film editors. They have a keen visual sense and an appreciation of flow and timing, but they also need to be highly organized to keep track of and manipulate what could be thousands of movie fragments that are being considered for use in a feature film. There will need to be two personas, one for the film editor in a large production doing only that job, and one for independent filmmakers who may write, direct, and edit their own personal movies.

Ecovative Design: Their customers will be corporations doing manufacturing and product distribution. They also have personas, which characterize their locations, the quantity and quality of their packing needs, the degree of automation involved, and any price considerations.

Airbnb: Consider a persona of users who are travelers. They almost certainly are adults. Some will be reasonably wealthy; others may be using Airbnb to do budget traveling. They will be somewhat adventurous, and probably keen to interact with their hosts and to obtain travel advice and guidance. They are most likely interested in short stays of less than a week, but some will seek extended durations. They will speak a variety of languages. There will be different personas for groups such as students looking for a place to crash while attending a conference and parents looking for family vacation rentals.

Airbnb hosts will also have personas. They will also be courageous, as they are opening their homes to strangers. They need their homes to look good and be very clean. They are sociable, keen to interact with their guests. They probably need the money that they can make by renting out rooms. They are keen to do well, as the ratings assigned to them by past guests will affect their ability to attract future guests.

Braze Mobility: Braze users are in wheelchairs. They typically are suffering from conditions such as stroke, vision loss or blindness, multiple sclerosis, cerebral palsy, spinal cord injury, traumatic brain injury, or dementia. Depending upon the condition, their navigational challenge may primarily be due to a mechanical condition, e.g., difficulties in swiveling one's head; a visual condition, e.g., difficulties in seeing obstacles; a hearing challenge, making it hard for them to know what is happening behind them; or a cognitive condition, e.g., recognizing that they are about to run into an obstacle. These different causes of their need for Braze's technology suggests that designers construct multiple personas for different kinds of users.

Drisit: There could be many kinds of Drisit users. One kind is a potential traveler, eager to see which exploration paths might be the most interesting. Other users are likely to be students in geography or history classes; teachers of such classes; travel agents and tour planners associated with tour companies, airlines, and cruise ships; parents of young people intending to take their first solo journeys; and individuals whose families are widely distributed around the world; and people who work for utilities who need to understand disruptions to their services caused by hurricanes and other natural disasters.

Two major challenges in constructing personas is determining what demographic and psychographic characteristics are salient and in gathering valid data with which to construct personas useful in design.

Use Cases

Personas encapsulate our assumptions about the people who will use our products. *Use cases* represent our best guess as to how those people will use the products. Product planners must devise use cases because they inform what functionality is needed for the kinds of users they expect. Use cases also guide developers and user experience designers in choosing what functions to emphasize and make easiest in the user interface to their products.

Adobe: Consider a use case for a film editor working on as part of a large production team with Premiere. Editors takes several clips and assembles them into candidate film sequences. They improve the constituent clips by trimming unneeded material, and experiment with sequences and arrangements of the clips, sometimes lengthening the feel of the whole by overlapping segments and applying special effects such as dissolves. Over time, candidate film sequences grow longer and longer. Editors will also construct multiple options for review by the director, producer, and possibly the actors. There is a need for clip and sequence labeling and organization tools.

SideFX: Considering the related technology of tools for creating animated digital media with SideFX software, use cases would be needed for the four market segments of movies and TV, game development, motion graphics, and virtual and augmented reality, as the design and production process may be different for each industry.

LinkedIn: There will be a variety of use cases for LinkedIn users depending upon their employment status (happy in one's job, considering leaving one's job, unemployed, student about to graduate), and whether their use of the app is driven by an immediate need for new job options or by long-term career-oriented social networking.

Airbnb: For most travelers finding accommodations, their use case has them specifying travel destinations, a range of dates for stays in that location, and attributes of the desired booking. They will usually include price and the number of beds and baths, but may also include amenities such as ocean view, parking, Wi-Fi, and facilities for eating and possibly cooking. In times of travel stress, there may be requirements dealing with cancellation privileges and cleaning. The use case will describe a clear and straightforward path for inputting this information, and for browsing among options, viewing photographs, requesting information, and ultimately making a booking. The use case must also deal with the possibility of changing a reservation, continuing to gather more information, communicating with the host before and during a stay, and submitting an evaluation of the host and the property after the stay.

Hosts of Airbnb properties use the software when setting up their home or apartment for future rentals, when responding to an inquiry from a potential guest before and during the guest's stay, and after the guest departs. The most time-consuming operations occur during setup, describing properties in detail and taking, organizing, and inputting appealing photographs. The most complex processes arise when they receive multiple conflicting requests for the same room on the same day and when they are answering questions from several different potential guests concurrently.

Beyond Meat: Product designers will have to consider the variety of ways in which people shop for, use, and prepare meat.

Blue Rock Therapeutics: Designers will have to develop use cases describing how medical materials manufactured from stem cells will be used in surgery and in pre-surgical and post-surgical treatment.

Braze Mobility: There are many different use cases for wheelchairs equipped with Braze sensors. One dimension of variability is the challenges faced by a senior wheelchair user, which could be motor, sensory—sight or hearing, cognitive, or some combination of these. Use of the wheelchair also depends upon location, indoors or outdoors, and the nature of the environment encountered, for example, empty or crowded with people, filled with few or many obstacles, smooth or rugged terrain, and the complexity of the desired path for the wheelchair.

Drisit: One use case for Drisit will describe individuals who desire to explore a specific site—country or city or area such as a coastline, or attraction such as castles in England or mountain ranges and peaks in New Zealand. Yet there may be other situations worth supporting with the app, such as a Surprise Me tour within a specific country or locale, such as islands in the South Pacific. An interesting technical challenge would be supporting concurrent exploration of an environment by several individuals on their own computers in different locations.

Developing use cases requires design imagination as well as a deep understanding of potential users of a technology. In all use cases, it is important to anticipate how usage will affect other stakeholders, such as neighbors of an Airbnb rental, other residents in a crowded nursing home who encounter a Braze wheelchair, or residents in a town being explored via a Drisit tour.

Market and Customer Research

Personas and use cases are helpful in product design because they inform the designers and product planners about the characteristics of their planned customers and suggest how these customers might use products to solve their problems and to alleviate their pain.

Effective product design and product management must also involve users in many other ways to ensure that the results meet user needs and are usable productively, comfortably, stress-free, enjoyably, safely, and with as few errors as possible. There are several ways to do this. We can watch and study how users currently work or learn or play without digital tools, or with the digital tools they currently use. This is called *ethnographic field research*. We can ask them what they think they need and about the likes and dislikes. This is often called *market research*. We can watch them work with prototypes of the tools we are planning to build, or the tools themselves. This is often called *user testing* or *usability testing* or *usability research* (I prefer the latter term).

The generic term encompassing all these processes is *user experience* research, which seeks to derive insights from users and about users working with technology to ensure that products are usable. In discussing this, I shall refer to activities in the life cycle of a product: *product planning*, *design*, *development*, *testing*, *refinement*, and *shipping*. These do not happen strictly in sequence and only once. Insights from developers during development, and later from testers or customers, will cause wise product teams to go back and do some redesign. This will happen repeatedly throughout the life cycle of a product.

Traditionally, product planners have sought to understand the marketplace by *market research*, which Investopedia defines as "the process of determining the viability of a new service or product through research conducted directly with potential customers. Market research allows a company to discover the target market and get opinions and other feedback from consumers about their interest in the product or service." The process is particularly valuable in yielding insights about the various market segments a firm is considering addressing.

Market research may be done in different ways. Surveys asking respondents to answer questions in writing may be done online. Alternatively, researchers may speak with respondents either over the phone or via teleconferencing. A year before starting MasterClass, its founder interviewed 12 people that he found over the internet to gain insights into needs and goals for online learning.

For the data to be valid, for it to contribute to a valid projection of a company's prospects, survey respondents must be chosen so that they are representative of the market being addressed. For example, a survey for Adobe premiere would address a variety of film editors working on movies and for television, and in different settings such as large studios and small independent production houses. A survey for Airbnb trying to understand who might want to rent their properties must include appropriate proportions of men and women, old and young people, and individuals of different racial and socio-economic backgrounds who might plausibly be traveling. It must also include people wanting to travel for different reasons and with different expectations in a rental. Questions must be phrased neutrally, that is, in such a way as to not lead the respondent to a particular answer.

In the same way that new products need to be tested prior to release (discussed below), sets of questions to be used in market research need to be tested to ensure that they are clear and that they are not biased in a manner that will skew the results.

Another method of validating an idea for a product for a market segment is to enter and study the life and culture of prospective customers in this segment. This is known as *ethnographic field research*. Ethnography is qualitative research that involves immersing yourself in a particular community or organization to observe behavior and interactions up close and personal. The term "field research" is used because the research is done in the field, in the environment in which prospective users work or learn or play. Ethnographic research can be "participant", in cases where the researcher becomes an active member of the community, and "non-participant", when the researcher is only a passive observer.

An ethnographic researcher trying to understand prospective customers for Adobe Premiere would live in a film editor's office and surrounding work environment for several days, asking lots of questions and observing how the editor uses his or her tools and collaborates with other professionals working on the production. Ideally, the researcher would do this in several environments, both in movie and TV production, and in large studio and small independent productions.

Brian Chesky of Airbnb recounted his actual experience in a "Masters of Scale" interview with Reid Hoffman, founding CEO of LinkedIn. In a meeting with Paul Graham, co-founder of Y Combinator, Brian was asked "Where's your business?" He replied that they were then mostly in New York city. Graham then said: "Go to your users. Get to know them." As a result, the co-founders of Airbnb commuted from California to New York for months. The knocked on doors, invited themselves in, photographed homes (for the web site), and even slept on the couches of their hosts. They learned what the hosts liked and didn't like, what concerns they had, and what needed to be in their profiles and in the profiles of prospective guests, and what might constitute a quality user experience for guests. This voluminous, ecologically valid user feedback informed the design and redesign of the Airbnb experience.

Pooja Viswanathan, founder of Braze Mobility, worked on smart wheelchairs for a decade before founding the firm. In doing so, she was able to enter and observe environments with wheelchairs such as hospitals and long-term care homes and observe and speak with wheelchair users who have a variety of challenges, as well as with staff caring for and assisting people in wheelchairs. She has admitted that to her detriment she had done too little of this during her PhD work, but that such immersion in the life of her prospective users during her postdoctoral work informed and guided subsequent strategic decisions and product design once she had formed her company.

As with surveys, it is critical that the researcher not let his or her behavior and the kinds of questions asked be influenced and distorted by preconceptions. Ethnographic research is particularly useful in the construction of personas, the abstractions summarizing the kinds of people and potential technology users encountered in the study and the use cases, what has been learned about how people might use an envisioned technology. It is also useful for understanding how individuals collaborate with other stakeholders in working, learning, or playing, and in their use of technology.

Despite the absolute necessity for understanding potential users and soliciting their input, it sometimes becomes compelling to go beyond what users can anticipate. Steve Jobs had visions of futures that went far beyond his customers' conceptions. The CEO of Caseware International has noted that there were cases that clients said that they wanted something, but he and his designers could imagine something that was even better. All of this can only happen with a deep understanding of your market and your users.

Prototypes and Usability Research

Another set of techniques used early in the software product design phase is the construction of *prototypes* having the essential features of the envisioned product.

Prototypes are answers to the question: How can prospective users give product designers feedback on their designs before the product has been built and is in a form that can be used? The answer is to construct something that can be built in 1/10th or 1/100th or 1/1000th the time, an artifact that portrays essential features of the envisioned product but is actually "smoke and mirrors". In other words, it gives a sense of what is intended, sufficient to elicit useful comments, but it does not need to work comprehensively in as many ways as would be required for the ultimate product.

Hardware prototypes constructed out of wood were used as early as the 1940s by the Polaroid Corporation to experiment with possible designs of their cameras prior to design and construction. Now, such as experimentation is done both with physical models and with interactive three-dimensional computer graphics.

Software prototypes are constructed in with traditional drawing tools and in special prototyping environments which can be used by non-programmers such as visual designers. They can be *low fidelity*, digital *wireframes* or paper prototypes only conveying the intended look of a screen and what capabilities it provides, or they can be *high fidelity*, interactive software also conveying the feel of the interface, allowing prospective users the ability to carry out a few commands as they would with the ultimate system. How much fidelity you need depends upon your goals. Paper prototypes may suffice to get feedback on the product concept, whereas high-fidelity prototypes will be required for *usability research*. There are available and easily accessible software tools such as Figma for creating medium- and high-fidelity prototypes of interactive software.

In usability research, investigators conducting tests will explain the purpose of the prototype and ask participants to work with it. Participants will be asked whether they understand what is on the screens and what is the purpose of each function accessible from a screen. Researchers will observe how participants react to the prototype, how they navigate between screens, what they able to do and where they seem blocked, what is easy and what is hard, what errors they make, and what misconceptions arise. Participants will be asked to *think aloud* to help researchers better appreciate what users understand and do not understand and why misunderstandings arise.

Software prototypes are sometimes constructed using a *Wizard of Oz* strategy. For example, this was used in the postdoctoral work of Braze Mobility's founder Dr. Pooja Vizwanathan. Some of the logic controlling the "smart wheelchair" was not provided by an algorithm, but by a smart person (behind a metaphorical "curtain"). enabling her to experiment with the kinds of control and feedback that would be essential features of various methods of collision avoidance. Her learnings informed the design of functionality and user interface for smart wheelchair sensors and then guided her programming of the desired functionality.

Improving designs by building and testing prototypes before building products helps avoid situations in which startups only become aware of many problems after they have invested hundreds of thousands of dollars in product development and have shipped the product. Mistakes caught early cost a fraction of what they would cost after most development is complete, and a tiny fraction of what they would cost in dollars, time, and reputation after the products are shipping. Prototypes are also useful in communicating the essence of your product concept to marketing and sales staff and to prospective employees, strategic partners, and investors.

Testing of usability and validation that your design seems to be viable must continue during product development. Large product companies have usability labs for this purpose. Such labs have one-way mirrors for observing participants, video cameras for recording participant behavior for later analysis, and monitoring software to enable analysis of phenomena such as the time it takes a participant to complete a task. Yet none of this is necessary, as an acceptable usability test can be done by sitting next to participants, timing things with a stopwatch, observing what they do, and asking questions.

Modern software development practices such as *agile development* insist that developers create frequent builds of their software, in some cases as often as every day. This has two benefits. It ensures that bugs will be found as quickly as possible. It also allows usability testing to be done as frequently as warranted given the complexity of the software and the degree to which past testing has uncovered problems.

Usability problems can also be discovered by thinking, analysis, and discussion by developers. *Structured walkthroughs* allow designers and developers to walk through the logic of use of their software and discover flaws in the interface during development, not after the product ships once they are much more difficult and costly to fix.

Products not comprised of digital technologies also require usability research. For pharmaceuticals, rigorous testing in clinical trials with increasing numbers of participants is required (see Case L. Antibe Therapeutics for a more detailed description).

Quality Assurance and Dogfooding

Once a product is relatively stable, more rigorous testing must be done, which is known as *quality assurance* (QA). One mistake many startups make is to skimp on QA because of a perceived need to release a first or follow-on product quickly for marketing or sales reasons. The result can be that the product ships with major errors—it does not do what it is supposed to do—or major usability problems—it is difficult to learn or use, and users get confused or get stuck. The embarrassment, the poor press, the critical social media comments, and the bad word of mouth can seriously damage a startup.

Testing of software has become a disciplined process. It happens in stages. Initial testing is done by the developers themselves, supported by testing professionals dedicated to that product. *Alpha testing* is then done by specific individuals within the firm who are not the developers or the formal testers. *Dogfooding* is also done by everyone else in the firm. Finally, *beta testing* is then done by friendly customers or people eager to obtain the product. Microsoft typically has hundreds of thousands of customers beta testing their flagship products.

Dogfooding is essential. Employees in the firm must use a product themselves, thereby gaining a deeper understanding of its strengths and weaknesses. This is known in the software industry as "eating your own dog food". Airbnb employees must use their system to book places to stay, Beyond Meat staff must sample and use their products in a variety of ways, employees within Braze must drive wheelchairs around their office while pretending to see or hear poorly or to be easily confused, and Drisit staff must explore the world with their app.

Customer Success

Finally, you are shipping your product. Since that is when you learn things that likely did not become apparent in-house, you want to ship as soon as you have a product that mostly works and does not have known killer bugs. Venture capitalist and founder of LinkedIn Reid Hoffman has been known to say: "If you're not embarrassed by your first product, you've released too late." I discuss what makes a shippable product further in Chap. 5.

Validation increases as sales and usage increase. Yet most firms will not be satisfied with early sales, as *going viral*—an exponential growth in sales—is rarely achieved. What are you doing right or what can be improved? Your customers are all over the world. How can you learn from their experiences?

One answer is another form of testing, different from that described above with prototypes. If you need to decide if a new feature is worth introducing, you can introduce with a significant subset of your users, for example, 10%, and through instrumentation and surveys understand if customers will use it and what their experience is. You then use this information to decide whether to add it to the product or not.

A variation of this is known as *A/B testing*. This applies particular to proposed user interface enhancements. Screen and menu organization is one of the best examples. You add code to your software implementing 2–4 different new interfaces and activate each version for a small percentage of your users. Through instrumentation and surveys, you learn which is preferred, which enable work most efficiently, and which is the most error-prone. This understanding guides you in determining which to introduce into your software's next version.

Another answer lies in responsive and empathic and inquisitive *customer support*, which now often labeled *customer success*. Yet many firms now have decided that this is a cost center that needs to be minimized. They list no phone number for their company on their web site. (There is now a website, gethuman.com, dedicated to make these hidden numbers known.) Their "Help" consists of providing a set of Frequently Asked Questions (FAQs) and Answers, which they try in good faith to make useful. These FAQs and bots may have seemingly plausible answers but an inadequate understanding of questions and the problems motivating the questions. Some FAQs will even include screen shots to convey greater clarity in their explanations, or video screen captures showing users carrying out tasks properly.

If they allow users to speak to human support personnel, it is typically through text chat. Such policies are penny-wise and pound-foolish. There are two approaches that I recommend, to be used in part based on the nature of your product and your customers.

One method is to orchestrate and nurture a user community network that will help one another with problems. Individuals in such communities often demonstrate incredible knowledge and commitment to helping others. Do understand that this engagement comes with a justifiable expectation that individuals investing significant time in helping people with a product are part of the team and that their reasonable suggestions for additions and modifications to the product will be seriously considered.

The other method, which I recommend even if the bulk of support is done by the user community. It is to create and nurture a support department staffed by knowledgeable, empathic, and accessible humans who will speak to puzzled or frustrated users. There are many benefits from this approach. Your users are much happier, and they speak well of you. Even more important is that the conversations are a source of market and product intelligence, which is why the support team also needs to be inquisitive. You will learn what features customers are using and what missing features they need and what they like and dislike and what problems they are having. In helping them overcome their difficulties, you will understand these more deeply.

This knowledge can be transmitted to the product management team and to the designers, which they will use to improve further versions of the product and even to conceive ideas for new products. It is important that the *voice of the customer* be heard at the highest levels of the company, as for example can be achieved by having the head of support be on the senior management team reporting directly to the CEO.

In the earliest days of a startup, the developers should do the support, and hence the learning, although you will eventually transition to a dedicated support team. This will be costly, but many firms will be able to use premium support as a cost center to reduce the net cost.

Clever use of technology is essential for effective customer support. Personnel helping puzzled users can do a better job if they can link their computers to customer devices so they can see what customers are doing and what they are seeing. Also, product software should be instrumented to gather anonymized data revealing which features are used and which are not used, how long tasks take, and what kinds of errors are made.

I now repeat myself because of the importance of the following point: Do not be misled by my account of the development process as proceeding directly from planning to design to development to testing to shipping. Life does not proceed in such a linear manner. Firms must commit to *iterative design* and development and improvement in which at any stage one may need to go back and redo steps, even to the extent of a fundamental redesign of functionality or user interface.

In summary, this chapter has presented a variety of user experience methods which increase the likelihood of product success and customer satisfaction and which can be applied throughout the development life cycle all the way to product shipment. Once the product starts shipping, even if it is not perfect ("the perfect is the enemy of the good"), it becomes even more important that the company and the product have a clear and appealing identity, which is the topic of Chap. 4, and where the concepts of a business model and a go-to-market plan become relevant, which are the topics of Chap. 5.

J. Wikipedia

Wikipedia was started in 2001. Co-founder and leader of the project, Jimmy Wales, sought to create "a world in which every single person on the planet was given free access to the sum of all human knowledge". There are now over 59 million articles, 6.5 million of them in English, in over 300 languages. Accessed by nearly 2 billion unique viewers each month, it has become the dominant general encyclopaedic resource in the world and the 7th most popular website.

Wikipedia's content is created by its community of users and readers. There are now over 103 million registered users, and over 280,000 active editors. Most content creators do not have special credentials but are just ordinary people with a passion for knowledge. Although the original goal of allowing text to be entered by anyone has had to be modified due to disputes over accuracy and occasionally even vandalism, various scholarly analyses have given Wikipedia high marks for quality.

Wikipedia has dealt reasonably well with the problem that there are sometimes passionate disagreements as to what is the truth and what is knowledge and what is neither. We aim "not for the truth with a capital T, but for consensus", said Wales, in which people who disagree can at least reach a consensus on the essence of their dispute. Nonetheless, there are still difficulties in the process of dealing with controversial material, and the approval process has become sluggish, with stated turnaround times of at least four months.

Another negative development has been the censoring of access to Wikipedia by over a dozen countries including China, Cuba, Iran, Russia, Syria, and Turkey.

K. LinkedIn

Internet entrepreneur Reid Hoffman wanted to change the world in a big way. He had originally planned to be an academic, but he realized four months into a master's degree program in philosophy at Oxford that there were better ways to make an impact. His chosen path was through becoming a software entrepreneur.

In 2003, he co-founded LinkedIn (LI) as a website targeted at the niche market of professionals eager to expand their career prospects or find new employment opportunities through social networking. LinkedIn is an excellent example of a firm resisting the urge to compete in larger markets but instead establishing, solidifying, and defending a leadership position in a niche that was ample enough to build a huge business. By 2021, LI's revenues had grown to over $10 billion a year; its membership in February of 2022 was 810 million individuals.

Hoffman's path to the founding of LI is interesting. He had taken enough programming courses to realize that he was not a brilliant coder, so he deliberately tried to learn about how to conceptualize and manage products and how to market, distribute, and sell them. In other words, he architected a curriculum of jobs that would teach him how to become a successful software entrepreneur. This meant a succession of jobs, first with Apple's eWorld, then with Fujitsu, then with his own unsuccessful startup SocialNet.com, and finally as COO and board member of Paypal in its formative early growth phase leading to enormous success. With SocialNet.com, he has said, he would review at the end of most weeks what he did not know but needed to know at the beginning of those weeks. He has stressed the importance of being a quick learner.

Hoffman replaced himself with experienced social networking executive Jeff Weiner in 2008, who orchestrated a mutually beneficial sale of LinkedIn to Microsoft in 2016 (see Chap. 7).

L. Antibe Therapeutics

Since getting his PhD from the University of Toronto in 1985, and primarily as a Professor of Pharmacology and Therapeutics at the University of Calgary, John L. Wallace has devoted his career to investigating the therapeutic properties of hydrogen sulfide. Encouraged by a 2002 scientific paper describing hydrogen sulfide's role in nerve signaling and homeostasis in the gastrointestinal and other body systems, Wallace has sought to demonstrate and understand the positive impacts of hydrogen sulfide in maintaining digestive system health and as an anti-inflammatory able to reduce pain. He has published over 500 peer-reviewed papers and is among the top 0.5% of biomedical scientists worldwide as measured by citations (over 42,000).

In 2005, Wallace brought in Dan Legault as CEO to help him build Antibe Therapeutics, with the stated goal of turning his research into pain and anti-inflammatory medications using gaseous hydrogen sulfide. Legault, originally a rescue pilot who had become a venture capitalist with a conscience and a commitment to health care, had served on the board of the leading Canadian health benefits administration company and of Save the Children International. He had also been responsible for managing small teams to build businesses in tech and other sectors.

The Company was initially funded by private sources before deciding in 2013 to list on the Toronto Venture Exchange. Access to VC funding has been a perennial issue for biotech ventures based in Canada.

It can take 10 to 25 years for drug discovery to yield a product. Laboratory scientific research, often aided by software, gives indications that certain molecules might have beneficial medical effects. These molecules must then be embedded in other substances to yield a potential pharmaceutical. This is followed by several years of tests on animals, with possible improvement of the compounds between tests. There then follow years of clinical trials, first Phase One on a few participants, then Phase Two typically on hundreds, and finally Phase III on a much larger group. If health problems or other side effects are detected in any phase, the company may need to modify the compound. Only after all phases are completed to the satisfaction of the U.S. Food and Drug Administration will it be approved for sale.

Identity

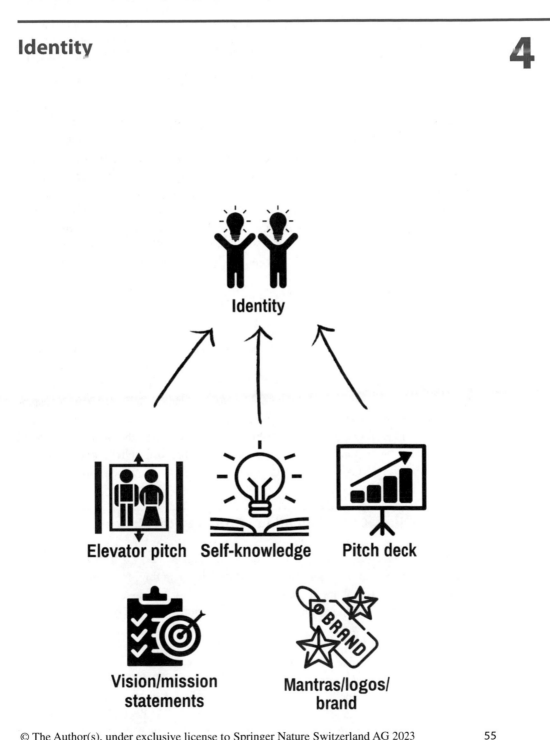

> On January 24th, Apple Computer will introduce
> Macintosh. And you'll see why 1984 won't be like
> "1984"
>
> Superbowl commercial introducing the
> Apple Macintosh (1984).

In the second half of the 1984 Superbowl, fans who had not left their seats to get another round of cold beers saw a remarkable ad. There are soldiers marching. There are rows and rows of ashen dazed people in prison garb, all staring at a screen. The huge ominous face of Big Brother is on the screen. His voice drones on, ominously and monotonously. Suddenly, we see a young girl, an athlete, running up the aisle between the citizens. She is carrying a large hammer. She sprints, as if afraid that she might be stopped. She stops, rotates twice while raising the hammer, and flings it at the screen, striking at the head of Big Brother and shattering the screen. Big Brother is of course the symbol of totalitarian control, aided by technology, and perhaps reminiscent of IBM— Big Blue—at that time the major supplier of personal computers. The text of the final screen reads, "On January 24th, Apple Computer will introduce Macintosh. And you'll see why 1984 won't be like "1984".

This chapter discusses the major elements of a startup's identity and how its thoughtful and creative design can support the launch of a new product and a firm's ongoing success. I shall discuss the need for the founders to know themselves and their strengths, goals, and values, and for them to translate these into effective descriptions of corporate identity such as *mission and vision statements*, *mantras*, and *logos,* all aimed at establishing a compelling *brand*. I shall also present two important methods of conveying a firm's identify and its promise of success—the *elevator pitch* and the *pitch deck*. I shall stress that a clear understanding of who you are and what you believe is essential for making and delivering on commitments to yourself, to customers, and to society.

Strengths, Values, Goals, Founding Teams, Ethical Choices

A prerequisite for creating a startup is self-knowledge and awareness. Knowing your values and your strengths and weaknesses are essential for planning, for achieving growth and profitability, and for honest relationships with employees, investors, and customers. Thinking through your goals and your lifestyle options is a prerequisite for sensibly making the decision to create and run a startup.

Readers of this book will see that I am very enthusiastic about entrepreneurship. Yet, based on personal experience, it is critical that you think through not only external factors that will determine success or failure of a new venture such as problem, opportunity, solution, market, and competition, but internal factors about you as a person and your internal resources and situation in life. Are you prepared, intellectually and emotionally, for embarking on a difficult journey which may last for many years? Can you do so without jeopardizing your marriage or partnership or the care and welfare of your family? Does your life partner understand the risks and stresses of the journey and support it without major reservations? Is **now** the time to do this, or would you be better served by waiting for the next good idea, perhaps in two or three years? Before embarking on a startup. consider dispassionately that choice in comparison to your other opportunities at the time, such as taking an interesting high-paying job where you can learn from watching their senior management in action.

Perhaps the most critical factor in deciding if you should create a startup now is the composition of the founding team. You will be overwhelmed with things to do, decisions, opportunities, and crises. Success rarely comes without years of hard work. Success rarely comes to individual entrepreneurs. You need founding partners. I discuss plausible configurations of startup management teams in Chap. 7. All founders must be willing to work very hard, be committed to the cause, and be honest and open with one another. Yor must be as confident of their integrity and other personal qualities as you would want to be in choosing a life partner. Having significant and positive prior experience(s) in working together is highly desirable.

Founders have values. Ethical startup formation should be consistent with these values. If you abhor war, and think that almost all wars are unjust, do not work on smart weapons; instead, work on systems to monitor adherence to treaties and ceasefires. If you value privacy, do not develop surveillance technology; instead work on privacy by design. If you feel that social media have turned evil with hate speech and disinformation, do not develop apps that rely upon synergies with Facebook; rather, work on new secure private social networks. If you are concerned about ransomware, consider a career in cybersecurity. If you believe that moving your labour force offshore to increase your profits at the expense of the health of your country or your region is unethical, commit to expansion at home, in the way Shaw Industries has grown to 22,000 employees, almost all of them in the U.S., and mostly in its hometown of Dalton, Georgia.

Such choices are not always easy, because ethics is rarely black-and-white. Consider, for example, facial recognition. This can be used to guard against terrorist acts and ensure justice for a crime, but also for surveillance by totalitarian governments and in a racially biased manner by the police. Even the firm that has gotten the worst press for evil uses of premature use of its facial recognition software—Clearview—recently applied its technology on behalf of Ukraine in its fight for survival against Russia. The current state of such technology has been widely criticized because it is not yet reliable. Accuracy is worse if the images are not of white males. IBM stopped working on it. Microsoft issued a call for government regulation and responsible industry measures to deal with issues such as bias and discrimination, invasion of privacy, and protecting democratic freedoms and human rights. In June of 2022, Microsoft updated its Responsible AI framework and restricted access to its facial recognition software and especially to its emotion detection tool because it did not reliably recognize emotions, especially when applied across nationalities, cultures, and races.

Technology itself is neither good nor bad, but the ethical startup founder will be sensitive to the current balance of good and bad. Think ahead before you take a job and start coding. If you feel that your work is more likely to be used for evil, you should factor this realization into your choice of work. Look for technical and startup opportunities where you can solve real problems, reduce customer pain, and make a better world.

Mission and Vision Statements

The Macintosh was announced as "the computer for the rest of us". This was very attractive to the 98% of people who were not part of the traditional tech priesthood, i.e., those who could type complex command lines to operate their machines. QLess advertises "No lines. Smarter operations. Happier customers/students/citizens/patients/…" to promote its app, which allows people to wait in virtual lines as opposed to standing in physical lines. Nuula's web site asserts "Your business at your fingertips".

These examples are statements of identity, asserting why a company exists and what the firm and its products do well. Two ways of describing the essence of a venture are *vision statements*, describing their end goals, and *mission statements*, describing what they need to do to achieve the goals. Here are some examples:

SideFX: *"For over thirty years, SideFX has been providing artists with procedural 3D animation and visual effects tools designed to create the highest-quality cinematic results. We are passionate about what our customers do because our roots are in production, both as artists and as pioneering technical innovators."*

Caseware International: *"We are leading the digital transformation of the audit and accounting industry. Our technology and people create impactful experiences that empower our users to deliver more value with trusted, actionable insights. Together, we will accelerate economic growth worldwide."*

Desire2Learn: *"Personalize learning, increase engagement, and help learners achieve more than they imagined possible. D2L offers flexible and robust learning solutions for every stage of life, from the earliest days of school to higher education and the working world."*

Antibe Therapeutics: *"Commercializing a breakthrough advance in inflammation science … Antibe Therapeutics is a publicly traded biotechnology company exploiting groundbreaking advances in inflammation science."*

Ecovative Design: *"We grow better materials. Ecovative makes it possible to enjoy the products humans need while sustaining and enhancing life on planet earth."*

Winterlight Labs: *"Monitoring cognitive impairment through speech. Winterlight has developed a tablet-based assessment that is fast, objective, and stress-free. By analyzing speech alone, we can detect cognitive impairment associated with dementia and mental illness. Our assessment can be used in life science research, senior care, and clinical settings."*

New York Times: *"We seek the truth and help people understand the world."*

Nuula: *"Nuula is a revolutionary application empowering small business owners to succeed. Nuula provides instant access to critical business metrics anytime, anywhere, plus a flexible Line of Credit that allows small business owners to access funds for just $1/day per thousand borrowed."*

Careful reading of these examples shows that they describe why the company exists, what kinds of products it offers, and why they are important, i.e., meeting the needs of some humans or businesses. They describe who these human customers are. In Caseware's case, the statement makes clear that their market is worldwide. Each statement asserts its purpose and goals and speaks of pride, quality, and excellence.

Alex Backer, founder of Drisit, recently provided a striking example of a startup vision. Noting that Alexander Graham Bell's invention of the telephone allowed people to hear anywhere, he suggested that Drisit's shared economy for drones would allow people to see everywhere!

Names, Mantras, Logos, and Brands

Other critical elements of a firm's identity is its name, its mantra, and its logo, for the company and possibly for individual key products.

The *name* must be concise, memorable, and evocative. Apple was not a particularly good name, but it spoke of human values in an exploding personal computer industry. Microsoft was a great choice for a name. SideFX was excellent because of its nearness to "special effects". Desire2Learn was imaginative with a lovely connotation. Wikipedia, Twitter, Qless, Ecovative Design, Chess.com, Beyond Meat, MasterClass, and Dsrisit were brilliant choices, as was Wordle as the name for a word game created by a man named Wardle. In general, however, it should never include the name of the founder unless s/he is as well-known as Elon Musk.

A *mantra* is short memorable phrase that evokes a company's vision and mission. An excellent example is the original description of the Apple Macintosh, "the computer for the rest of us". Other examples are:

Imax: "The world's most immersive movie experience"
LinkedIn: "Welcome to your professional community"
Airbnb: "Book homes from local hosts"
Chess.com: "Play chess for free on the #1 site"
New York Times: "All the news that's fit to print"
Canva.com: "Online design made easy"
MasterClass: "Learn from the world's best"
Blue Rock Therapeutics: "An entirely new generation of authentic cellular medicines"
Braze Mobility: "Blind spot sensors for wheelchairs"
Drisit: "Choose your adventure" and "Reimagining being there"
Nuula: "Your business at your fingertips" and $1 per day for a loan".

A *logo* is a graphical mantra. Here are some good examples from our case study firms. Go to their web sites to look at the logos.

We are all familiar with Apple's stylish apple with a bite in it. Microsoft's logo evokes its Windows operating system. Twitter represents itself with a graceful tweeting bird.

Braze Mobility's clever logo is elegantly designed to give the sense of a human inside a B possibly on some sort of chair but also leaping forward. Winterlight Labs is represented by a night scene echoing its name. Antibe Therapeutics displays an animation that includes the phrase "Targeting inflammation".

Put all together, these elements of a firm's description and its offerings of products and services define who it is and why it exists. Its identity can be bolstered by intellectual property protection, as for example Beyond Meat has done with its aggressive program of filing for trademarks, and some of our case study firms have done with patents.

All of this in synergistic combination become a *brand*. Here are some examples:

Apple co-founder and two-time CEO Steve Jobs's drive, technical intuition, design sense, and market savvy fueled continual innovation until his early death in 2011. Apple moved from being a shaky industry outsider to the inventor and skillful marketer of the most important breakthrough products of the high-tech industry—the iPhone smartphone, the iPad tablet, and the Apple (digital) Watch, as well as marketing innovations such as tracks of music sold for 99 cents from iTunes, the app store, and elegant retail stores for the sales and support of Apple products. The brand, supported by effective and elegant graphic, industrial, and media design speaks of quality and usability, and of being high-end, designed for knowledgeable and wealthy customers.

Microsoft has been the leader in software for microcomputers and personal computers since the earliest days of the industry in the mid-70s. It has systematically expanded its product offerings to include office productivity tools, cloud services, tablets, and gaming hardware and software, among a host of products internally developed or obtained through acquisitions. Unlike Apple, which has been a proprietary walled garden, Microsoft has positioned itself as the cornerstone of the vast Windows-compatible computer marketplace. Its has established a brand speaking of a strong developer community, universality, appeal to large corporations and consumers alike, and trust, the latter being somewhat paradoxical given its often-buggy software and susceptibility to security breaches.

Adobe, based on decades of innovation and strategic acquisitions, develops software to support document creation, publishing, photography, video creation and production, graphic design, illustration, marketing, and much more. All of this is now integrated into blockbuster products—the Adobe Creative Suite, also available as a Software as a Service (SaaS) offering called Adobe Creative Cloud. Its mantra of "creativity for all" coupled with its success in corporate, academic, and creative industry markets worldwide gives it a brand that speaks of productivity and trustworthiness, although the emergence of firms such as Canva shed doubt on the validity of the phrase "for all".

LinkedIn (from its website) "connects the world's professionals to make them more productive and successful. ... [with] 756 million members worldwide, including executives from every Fortune 500 company ... the world's largest professional net-work." LinkedIn's brand speaks to utility for professionals to aid them in professional networking with the goals of excelling in their current jobs and seeking new positions.

Beyond Meat's website states: "We believe there's a better way to feed our future. By shifting from animal to plant-based meat, we can positively affect the planet, the environment, the climate, and even ourselves. After all, the positive choices we make each day—no matter how small—can have a great impact on our world."

Elevator Pitches and Pitch Decks

Finally, there are the concepts of an *elevator pitch* and a *pitch deck*. Companies seeking capital in the 60s, 70s, and 80s wrote business plans, which were often 20- to 30-page documents, supported by appendices and spreadsheets encapsulating financial statements and forecasting models. Such business plans are no longer required nor desirable.

The best definition I have seen of an elevator pitch is "a short, pre-prepared speech that explains what your organization does, clearly and succinctly" (from www.mindtools.com). The word elevator is used because the assumption is that you find yourself in an elevator with a potential investor, and you must communicate your message before the elevator reaches the floor where the investor departs. If you are successful, you may get a meeting with the investor. Elevator pitches are also useful for introducing your company and/or your technology to prospective clients, partners, and employees.

An elevator pitch must include a description of who you are, what you do, your USP (unique selling proposition), and a call to action at the end.

Here are some examples that I have created. To illustrate the key ideas, I take liberties and make assumptions, both with respect to claims and the call to action.

Imax, the world's most immersive theatrical experience, is expanding into luxury home entertainment systems, click here to download our investment prospectus.

"Wikipedia, is a multilingual free online encyclopedia written and maintained by a community of volunteers through open collaboration and a wiki-based editing system." (from its Wikipedia page)

Ecovative Design allows your firm to pack and ship materials without inflicting further damage on the world's fragile environment, click below and fill in the form and we'll ship you free samples, also indicate if you would like us to phone you.

Chess.com is the world's foremost chess site, featuring instantaneous matches against appropriate opponents, retrospective game analyses, and many other tools for learning; join now and improve your game.

Beyond Meat allows you to do what you've dreamed of for years ... halt cruelty to livestock and enhance the environment without giving up the tastes you love, click here to order free samples ... chicken, burgers, sausages, and more.

Drisit is the world's first online platform for visits by drone, allowing the booking of a drone anywhere in the world, control of flight path and camera angle, and streaming of

your Drisit … you can now see anywhere in the world! We seek wizards to create magi-
cal experiences … apply today at ….

Nuula is a revolutionary application empowering small business owners to succeed
by providing instant access to critical business metrics plus a flexible Line of Credit for
access to emergency funding. Call today to understand our investment opportunity.

The final encapsulation of the essence of a start-up is the pitch deck. Ideally consisting of not more than 15 slides, this is what you need when you are presenting to a
venture capitalist. Variations will also be useful in presenting to family or angels for early-stage investment, and to potential customers and employees.

Although every company is different and pitch decks are used in different circumstances, I suggest the following this 13-slide organization as a starting point in designing
the deck:

1. Company name/mantra/mission/vision
2. Problem/opportunity and solution/market
3. Value proposition
4. Underlying magic (including IP)
5. Validation: use models
6. Validation: user experience research
7. Validation: competitive analysis
8. Business model (canvas)
9. Go-to-market plan
10. Managing finances and financial forecasts
11. Financing
12. Leadership, management, team
13. Call to action.

These topics are covered in the other 6 chapters. An example pitch deck for this book,
articulating its vision and special features, appears in the first appendix to this book, as
does a competitive analysis and an elevator pitch.

One key challenge is to create the pitch deck and the accompanying narrative so that
it tells a compelling story. This may require a novel way of organizing and presenting the
material. Pitch decks that sound like a succession of answers to questions such as "what
is your value proposition?" and "what is your go-to-market plan?" will bore the attendees and will surely fail.

In summary, successful entrepreneurial ventures know who they are and be able to
communicate it vigorously to customers, employees, and investors. Success relies upon
innovative solutions to big problems which convey significant value to customers, the
topic of Chap. 1, and on entering and dominating markets, backed by sufficient capital
and with the right team, which are the topics of Chaps. 5, 6, and 7.

M. Twitter

While still an undergraduate at New York University in 2006, Jack Dorsey originated and demonstrated the viability of the idea of using a social media platform as a vehicle for transmitting short bursts of information to large numbers of people via the internet. The 2020 Annual Report describes Twitter as "a global platform for public self-expression and conversation in real time."

Dorsey and a small team formed a startup to commercialize the idea. Their actions illustrate the choice of an evocative name and an appealing logo to help create a memorable brand, and the value of creating a new product category, as nobody had seen anything like this before. The best way Twitter had at the time of describing what they offered was "micro-blogging", although this is not a great way of conceptualizing it.

Twitter has been a huge success in terms of impact on the world, even though it is not viewed by analysts as a financial success. It now has about 200 million monthly users. 21% of U.S. adults use the service, as do 42% of Canadian adults with internet access. It functions as a vehicle for announcements and information transmission, and as a way for tweeters to exhibit their knowledge and build their brand.

Ethical behavior has long been a central preoccupation of Twitter. Social media firms, most notably Facebook (now Meta), have received worldwide condemnation for invasions of privacy and for failure to stop the torrent of hate speech, fake news, and other disinformation being disseminated on its platform. Twitter therefore has faced ongoing struggles to behave responsibly. Perhaps the most notable example is its weighing free speech against a safe and just society in dealing with the actions of its most famous tweeter, Donald Trump. Over the course of the last tumultuous year of the Trump presidency, it took repeated decisions to label some of Trump's tweets as "potentially misleading" and others as "glorifying violence", until it finally suspended the President's account and then permanently blocked his account.

While this book was in production, Twitter was purchased by Elon Musk. His statements and recent actions imperil both the firm's ethical stance and its very survival.

N. QLess

Alex Backer, my second cousin, has a Ph.D. from Cal Tech in Neuroscience, yet pains-taking detailed laboratory work was not suited to his energetic personality. He has become a prolific inventor and entrepreneur, including among his early ventures Whozat, which he termed "The People Search Engine", an excellent idea that was not commer-cially successful.

His next venture, QLess, has been dedicated to the elimination of the need for people to stand in lines for significant periods of time. The idea is simple. Instead of standing in a physical line, you enter a virtual line. You are summoned on your mobile phone with messages such as "you will reach the head of the line in 30 min", "we are sorry, but your entrance has been delayed by an additional 10 min", and "please go to the check-in counter in 5 min or you will have lost your place in the line". The software can also be accessed from desktop and mobile computers.

QLess has been successful in a variety of industries including amusement parks and event venues; colleges and universities; hospitals and other healthcare institutions; gov-ernments at various levels such as cities and counties; retail outlets of various sizes, especially during times of sales; and banks. Use of the software provides benefits to cus-tomers by removing the need to stand in long lines for extended periods and to institu-tions and companies by avoiding congestion, increasing customer good will, and in some cases allowing customers to spend additional money while they are waiting.

Alex left QLess in 2020 and launched Drisit in 2021. Drisit is discussed as example venture Y.

O. Ecovative Design

Rensselaer Polytechnic Institute (RPI) students Eben Bayer and Gavin McIntyre took a class called The Inventor's Studio taught by Burt Swersey in the fall term of their senior year. Their ideas were deemed impractical, but Eben brought to one of the last classes a solid white block of perlite particles that had surprisingly and mysteriously been glued together from wood chips by mycelium strands originating from mushrooms. (I discussed at the beginning of Chapter Two the origins of the idea in a recollection of Eben's from his days on the family farm.) Instructor Bert Swersy found this promising and suggested that they take the class again the following semester, and that they work on improving the ability of mushrooms to knit together waste materials to form insulation and packaging in a form that is totally biogradable.

Ecovative became a viable startup because the wood chips can be replaced with agricultural waste such as chopped-up corn stalks and husks and produce solids that imitate Styrofoam in form, function, and cost. Ecovative's product is biodegradable and hence not as destructive to the environment as Styrofoam, which litters the ground, fills waste dumps, pollutes the oceans, and accumulates in the digestive systems of animals.

The origins of the company are equally fascinating. After Gavin and Eben graduated, Burt kept phoning them, insisting that the pair return to the campus. He offered use of campus resources and a personal investment out of his retirement savings. Eventually, the two turned down full-time jobs, returned to campus into space in the campus startup incubator, patented their invention, and founded Ecovative Design in 2007. They were aided by several grants of funding from invention and entrepreneurship challenges followed by a 2010 research grant from the National Science Foundation and finally venture capital investments starting in 2011.

Business Models

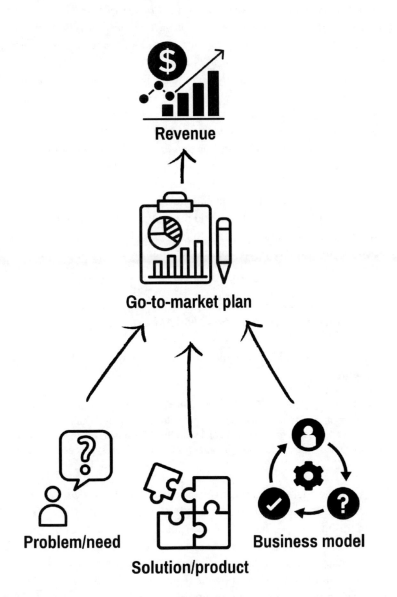

R. Baecker, *Ethical Tech Startup Guide*, Synthesis Lectures on
Professionalism and Career Advancement for Scientists and Engineers,
https://doi.org/10.1007/978-3-031-18780-3_5

"If you have more money than brains, you should focus on outbound marketing. If you have more brains than money, you should focus on inbound marketing."

Guy Kawasaki, https://quotefancy.com/guy-kawasaki-quotes

In 2021, Josh Wardle, a Brooklyn, New York, software engineer, created a simple online word game for his partner as a token of his love. Players of Wordle are given six tries to guess a 5-letter word; on each try, they are told which guessed letters are in the correct position, which appear in the word but in different places, and which are not in the target word. Over a few months, the pair and members of his close family played, followed soon by his WhatsApp extended family. The game was so popular that he published it online in October. On Nov. 1, there were 90 players. By Jan. 2, 2022, there were over 300,000 players, in great part because allowing someone to notify a friend about the game by sharing a graphic of his or her solution success helped the game spread virally. It was written up in the New York Times on January 3. The paper bought the game later that month for an amount said to be in the "low seven figures".

A *business model* is a description of how a company uses people and resources to deliver products and services to customers and how it makes money. This chapter discusses the elements of a business model and how paying attention to them is essential for product success. I pay particular attention to *go-to-market plans*, product design, pricing, web sites, and to *digital marketing* approaches that encourage the *viral* growth of customers and revenue.

Business Models and the Business Model Canvas

Alexander Osterwalder and Yves Pigneur, in their classic 2010 book, *Business Model Generation*, write:

> ... a business model can best be described through nine building blocks that show the logic of how a company intends to make money. The nine blocks cover four main areas of a business: customers, offer, infrastructure, and financial viability. The business model is like a blueprint for a strategy to be implemented through organizational structures, processes, and systems.

A business's offer must be attractive based on a *value proposition* (Chap. 1). Infrastructure includes *key resources* including management and a team (Chap. 7), *key activities* (discussed in Chaps. 2–4 and this chapter), and *key partners* (Chap. 7). Customers are organized into *customer segments* (Chap. 1), which often are reached via *channels* (this chapter) and are always supported by *customer relationships* (this chapter). Financial viability is determined by the degree to which the *revenue streams* exceed the expenses determined by the *cost structure*. These may be diagrammed via the following *business model canvas*.

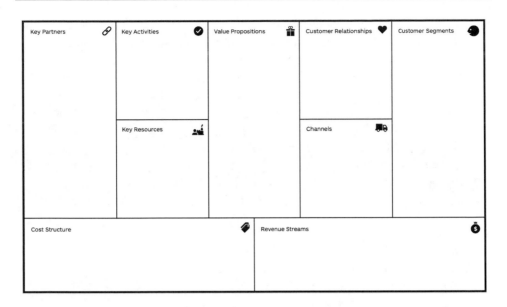

Most of these elements are used in a startup's go-to-market plan.

The **Apple** iPod/iTunes music product consisted of the iPod digital music player (2001) and the iTunes digital media service (2003), which add values to and encourages sales of the hardware. The iPod was later replaced by the iPhone.

Value Proposition: Individual songs became available at 99 cents each, an industry first. The combination of elegant hardware, software, reasonably priced abundant music, and a streamlined online store provide unparalleled convenience in searching, buying, and enjoying music. It therefore disrupted the music industry

Customer Segments: The mass market of popular music lovers

Channels: The product was available from apple.com, also from Apple stores and other retail outlets

Customer Relationships: Convenience and a compelling brand build customer loyalty

Key Resources: People, hardware, software, music content

Key Activities: Hardware and software and store interface design, lots of marketing

Key Partners: Record companies, artists

Cost Structure: People, manufacturing, sales and marketing, royalties to music publishers and artists

Revenue Streams: Hardware sales, some revenues from sales of music.

Twitter is a social medium that enables short (originally, 140 characters, now 280) statements of information and opinion (tweets) to be distributed widely over the internet.

Value Proposition: Twitter affords easy and free construction and sending of tweets, hence is a fluid platform for dissemination of the knowledge and views of politicians, thinkers, and opinion leaders. It thereby helps them build their brands and increase their influence

Customer Segments: The mass market of people interested in news and ideas

Channels: Twitter.com

Customer Relationships: Content moderation of dangerous and misleading information helps protect the integrity of the platform and is viewed as important by most users

Key Resources: People and software

Key Activities: Software development, content acquisition, sales, and marketing

Key Partners: Developers and commercial enterprises who find value in disseminating their data streams and archives

Cost Structure: People, sales, marketing, internet bandwidth

Revenue Streams: 86% of Twitter's revenue comes from advertising, the remaining 14% from data licensing. Advertising is attached to promoted tweets, promoted accounts (tweeters), and promoted topics (*trending*), which facilitate greater viewing of those tweets accompanied by ads. Twitter may be considering a subscription model for revenue.

The business model approach and the business model canvas is very valuable. It has had enormous success and is now used by many executives in rapidly changing industries and in many university business courses. Yet it does not capture all elements that determine entrepreneurial success. Here are some of its weaknesses.

It ignores the competitive landscape and its effect on corporate strategy. It is a static model and does not guide changes in strategy that happen over time. It does not speak to the processes of raising financing. It says little about the leadership and management skills required for success. It says little about the external environment encountered by a startup, and by the political processes and ethical issues that it may encounter.

This book covers all these important issues, both those included and those not included in the business model canvas approach.

Go-to-Market Plans

Implicit in a business model is its *go-to-market plan*. This plan encapsulates the strategy and tactics used by a startup to launch its product and move it forward in the marketplace. Strategic decisions include choice of a customer segment and the channels for reaching customers. Tactical decisions include obtaining and orchestrating the resources, activities, and partnerships required for a successful product launch. Marketing is especially important.

Apple Macintosh: The Macintosh was launched at the beginning of 1984. At that time, the annual fall COMDEX trade show in Las Vegas was the venue for showcasing digital technologies and for industry observers and journalists to learn about products they needed to cover. Yet the key to Apple's successful launch was the dramatic and memorable 1-min Super Bowl commercial discussed at the beginning of Chap. 4.

SideFX: Kim Davidson and Greg Hermanovic worked for HCR and then as key employees for Omnibus Computer Graphics, which in 1987 became the largest computer animation company in the world. But this was a financial overreach, and the firm soon had to declare bankruptcy, leaving Kim and Greg without a job. Cobbling together some money from family and friends, they put it a successful bid to the receiver for Omnibus's PRISMS source code (which they had helped to write) and incorporated Side Effects Software to develop, market, and sell computer animation software. Two customers of Omnibus became their first customers. The revenue from those contracts and other connections made at Omnibus helped fuel their growth and support the development of more advanced tools, culminating in the development of the award-winning Houdini package in 1996.

Chess.com: By the time chess.com launched in May 2007, thousands of chess players had found the site (in part because of it had the best name of all possible chess sites) and signed up to learn more. They then got more customers through word of mouth and because of excellent Search Engine Optimization (SEO, discussed below). But then luck rewarded their dedication and good work—an unexpected feature article in TechCrunch in July 2007, resulting in huge traffic to the site and an expedited path to success.

Braze Mobility: Although Pooja had learned a great deal in her doctoral and post-doctoral work, she launched Braze in 2016 with extensive stakeholder engagement, that is, making wheelchair users informal members of the design team. She learned that users did not want the wheelchairs to take total control, and that a smart wheelchair's best mode of operation would depend heavily on both the users' physical environment (home or institution) and their circle of care. There followed a planning process, integrating the results of research studies with information gleaned from customer interviews, to *pivot* the company concept from the originally envisioned highly automated chair to simpler spot sensors which could be attached to existing chairs. (I shall discuss other cases of pivoting in Chap. 7.)

The original concept had been to provide automatic speed correction to moving wheelchairs for people with cognitive impairments in long-term care as an initial niche market. The new concept was to mount visual spot sensors on a wheelchair and give audio and haptic feedback to provide wheelchair users with knowledge of obstacles and enable safer use and greater independence. The target market also was broadened to include individuals without cognitive impairments.

This strategy also had the enormous benefit that Braze's products would no longer be classified as medical devices with a need for lengthy and expensive regulatory approval. Hence Braze could develop the sensors and the control electronics, knowing that testing and validation with early adopters could be followed immediately with a vigorous go-to-market program of direct and indirect sales.

To go-to-market requires marketing. Marketing textbooks describe their discipline in terms of the *4Ps of marketing*—product, price, place, and promotion. *Product* encompasses the totality of a firm's offering for customers, which is much more than the software itself. *Price* is no longer a single quantity, as customers are asked to pay varying amounts at various times. *Place* is now typically the internet and the web; internet marketing to draw people to your place on the web is a complex discipline. Technology *promotion* has evolved significantly over the 75 years of the computer industry, although the basic need to capture and hold attention has not changed over that period. I shall add a fifth P—*partnerships*, as they are essential for tech startup success.

Products and the Minimal Viable Product

Harvard marketing Professor Theodore Levitt has described the *product* as consisting of 4 distinct possibilities: the *generic product*, the *expected product*, the *augmented product*, and the *potential product*. The generic product in a category is the minimum required to be able to say you are delivering something within that category. The expected product is larger, and therefore meets the customer's minimal expectations. Your expected product includes documentation. Support and training used to be part of the expected product, but this is no longer the case with many firms offering apps on the web or in an app store. The augmented product surpasses expectations and allows a firm to differentiate its offering from those of competitors. This now often includes support and training. The potential product encompasses not just what has been done, but what remains to be done, the gleam in the eye of savvy designers, developers, and marketers.

Consider Caseware as an example. Their minimal product consists of software tools to streamline the auditing process. Their expected product also includes documentation, training, data sets, case studies, and other aids to customers learning and mastering Caseware's technology. The augmented product includes their understanding of local customs and practices achieved by selling and supporting their software through a global network of distribution partners. The potential product has evolved from time to time, as their CEO has continually directed development starting from the core functionality towards the next technological frontier, first the desktop with a GUI, then the cloud, and most recently the incorporation of AI.

Programmers who do startups often make the mistake of thinking that their code is a product. Products comprise far more than algorithms realized as computer programs. Another key element is *data*, or *content*. For example, Chess.com's augmented product includes its collection of instructional materials for players at various levels, its different game types, and its large membership of chess players at all levels, thereby making it possible for a player to instantly find someone well matched as an opponent. Airbnb succeeds because of its large quantities of satisfied customers, individuals and families seeking to rent out extra space and those traveling seeking congenial accommodations. MasterClass's treasure is its library of quality recordings of classes taught by the top world experts in many fields.

Augmented products become richer and more valuable if capabilities can be added by clients and other firms. The way to do this is to include an *Application Program Interface* (API) in your product. This means that programmers working for your clients, guided by the clients' usage and production needs, or those working in other firms, guided by their product and marketing needs. can use your software to enhance their productivity and build new applications. APIs add significant value to your product and may be a direct or indirect source of revenue.

A great example is SideFX. Customers such as computer animation or special effects studios can create graphics and animations using the SideFX tools, which were built using algorithms realized by the firm's lower-level procedures or processes. Programmers working for the customer also have direct access to these same processes, which they can then adapt to create new capabilities and moving images. The flexibility and power of the SideFX API is one of the major ways that the firm differentiates itself from competitors.

A final comment about products deals with the issue of completeness and timing. I repeat Reid Hoffman's famous quote: "If you're not embarrassed by your first product, you've released too late." You will never satisfy all your expectations with a product release, and all releases will have some bugs. You must therefore avoid the temptation to include everything and to continue the process of adding more and more. It is important that your understanding of your customer and the market be nuanced enough so you can define the *minimum viable product* (MVP) that is sensible to ship. So long as you are releasing the system without major bugs that will be crippling for significant numbers of people, ship the MVP! Then learn from your customers' experience (as discussed in Chap. 3) and make the next version much, much better, including adding some of the functionality your developers are so keen to add. Keep in mind that the perfect is the enemy of the good!

Revenue Models and Pricing

The next P is price. Before discussing this, I must speak of *revenue models*. Some firms sell to other businesses (these are known as business to business, or B2B). Some sell to consumers (these are known as business to consumer, or B2C). SideFX, D2L, Ecovative Design, and Nuula are B2B. Chess.com, AirBnB, Beyond Meat, and MaserClass are B2C. LinkedIn and Twitter do both, as individuals seek professional advancement and spread their ideas on the platforms, but firms can also use them to attract employees and to build their brands.

Many B2B firms have their own *direct sales* force, which is used for large volume sales or high value B2B sales, with the price of an order often being tens of thousands to hundreds of millions of dollars. Yet firms have a choice between selling direct to their customers in this way or via a *channel*.

Channels are ways to get your product into the hands of customers indirectly, through the efforts of the direct sales force of another organization. Here are the major kinds of channels for software. For expensive programs, *sales agents* represent your software, often in distant geographies, and in return receive a commission on every sale. *Distributors* often sell your software directly but sometimes have agents working for them in more localized areas. Caseware International is a great example of a firm extending its reach internationally by partnering with distributors located in other countries that have strong accounting and software expertise and good knowledge of the local culture. Distributors make the sales and remit much of the income to the publisher of the software.

Certain kinds of "low level software", such as device drivers and data conversion software, are bunding in with some other firms's hardware (this is known as selling via an OEM, an *original equipment manufacturer*). Authors of such programs may only receive pennies for each sale, but they have little to do but fix bugs and keep the software current.

Given the ubiquity and reach of the internet and the web, most B2C software now uses as its primary distribution channels a *web site*. Apps are downloaded from that site via the internet. Publishers receives the full price, but they bear the full burden of getting potential purchasers to their web sites (discussed below).

The same categories apply to firms selling tangible items. For example, Braze Mobility does some sales direct, for example, to the U.S. Veterans Administration, which is a client with many hospitals all over the United States. Braze also has distributors and dealers in certain geographies.

Deciding on *price* is not easy. There are several approaches. I characterize them as competition-, value-, and cost-based. Most pricing is done with respect to what competitors charge, either aiming low to be a bargain or aiming high to signal quality. If it is possible to quantity the value that your product delivers, an effective sales rationale is that your product costs X and it saves the customer Y, where X is greater than Y, then you set the prices knowing that your product will only actually cost the customer X–Y. You also want to ensure that the costs associated with each sale do not exceed the price of the product, otherwise you will lose more, and you will lose even more money the more sales you get. You can lose money for a while, for a very long while if you are well-funded and you are building market share, but eventually viability depends upon economies of scale reducing your costs per sale so thar you are making money.

A major risk factor for a startup in launching a product in competition with a Big Tech firm such as Amazon, Apple, Facebook, Google, or Microsoft is that they have war chests of tens or hundreds of billions of dollars. They can afford to keep their prices low indefinitely while you slowly bleed to death financially. The same holds true if one of them decides they like your product idea and then launches a competitive product.

The numbers of people who could download apps from the internet numbers in the billions. Hence most consumer products are priced at no more than $20; some cost less than 1 dollar. However, there are three other ways in which software can be priced, *Software as a Service (SaaS)*, *freemium*, and *free*.

The idea of software as a service is that you pay every month as you use the product, rather than paying a much larger amount when you initially order the product. For example, MasterClass charges not for each class you take, but a monthly fee for access to all classes. There are three levels of annual memberships, with different prices based on the number of devices that can be concurrently accessing the content. Another good example of SaaS is Canva.

The idea of freemium software is that a minimal or restricted functionality version is free but that you must pay for more features. Those payments typically recur monthly. For example, LinkedIn's basic functionality is free, but a premium membership allows enhanced access via private messaging to people you want to meet. Chess.com is free to play, but a monthly charge unlocks the ability to get a detailed analysis of your games, showing you what you did well and where you made mistakes, and what you could have done instead that was better.

Finally, many apps are now free. Their vendors make money by having sponsors or by advertising. Vendors whose products cost money also generate additional revenue using ads. Here is an illustration of the key idea. When I am using LinkedIn, I occasionally see Ads about products and Invitations about events on the screen. The ads and invitations will be most useful and generate see the greatest revenue if they are keyed to the then visible content. In other words, an ad for an executive search firm would appear when a LinkedIn user is viewing information about a vacant position.

It is important to build a forecasting model (more on this in the next chapter) in which your customer and growth forecasts predict financial performance, and where you can update the model regularly based on actual results to yield evidence as to the viability of your pricing strategy.

The Places Known as Web Sites

The third P is *place*. With traditional businesses such as retail stores and restaurants, place was key. This was encapsulated by the slogan "location, location, location". Restaurants, for example, sought to be on streets that were: (1) near other restaurants; (2) frequented by shoppers; (3) having adequate inexpensive parking; and (4) near residential areas with people who have the inclination and means to go out to eat. Even in the early days of the computer industry, at trade shows like Comdex where media, large buyers, and software developers gathered to see the latest toys, firms like Microsoft, Apple, and Visicorp sought booth locations that would maximize visitors from both from those consciously looking for their booths and people wandering by serendipitously. They also sought to maximize their exposure in industry newspapers and magazines such as Wired.

The new "electronic storefronts" and "electronic trade show booths" are the website, Apple's App Store, and Google's Play Store. In the same way that a storefront must be eye-catching and attractive and communicate what is available within, websites and app store presences need to be designed to have similar attributes. Just as large stores need tools for wayfinding and navigation, so that you can find what you want efficiently, web sites need to ensure that you find what you want quickly without getting lost. In the same way that stores need to ensure that their goods meet customer needs, website and app store content must be appropriate and sufficient for realizing sales. Content must also be kept current, as electronic commerce proceeds at a breakneck speed.

A startup's website and app store presence is the place it occupies in cyberspace. They must be clean and attractive, containing goods and services to buy and media that enable customers to make informed decisions. These media can be traditional documentation, videos for persuasion and information, and access to knowledgeable sales and support personnel (more on this below). They must convey a sense of competence and presence. Above all, they must be findable by potential customers, which brings us to the fourth P, promotion.

Promotion and Internet Marketing

Promotion used to be done via postal mail, newspapers, magazines, radio, and television (recall the Apple Macintosh *1984* ad). Since the advent of the internet and the web in the early 90s, more and more attention and communications are being targeted via electronic media such as email, web sites, social media, and video streaming. Statista.com reports that the average time per day spent with traditional media versus social media in the U.S. since 2011 had been:

	2011 (min)	2021 (min)
Traditional media	453	318
Digital media	214	482

With the steady growth of digital and decline in traditional media, the two curves crossed sometime between 2017 and 2018.

The social media most used worldwide in 2021 were:

Facebook	2.9 billion active users
YouTube	2.3 billion
WhatsApp	2 billion

Instagram	1.4 billion
Messenger	1.3 billion
We Chat	1.25 billion
TikTok	1 billion

Social media differ primarily on their degree of support for various kinds of messages and the demographics of their users. Facebook is the giant, and its parent company, now known as Meta, also owns WhatsApp and Instagram. YouTube has for over 20 years been the largest for video content. Instagram is very strong on photographic content. WhatsApp is very strong internationally; it also supports messaging and synchronous communication. WeChat is similar, is dominant in China, and directly hosts lots of e-commerce. TikTok also originated in China, is optimized for the creation and hosting of short videos, is supplanting Instagram as the site for youth (which supplanted Facebook in the allegiance of that demographic a decade ago). TikTok also may soon overtake YouTube as the dominant site for video content.

Internet promotion relies upon the same basic principles as characterized traditional promotion, namely, addressing communicative, authoritative, attractively packaged advertising messages to the right target audience.

In some cases, the message is *pushed* to potential customers via email. The key to success is creating or purchasing a suitable mailing list. You must have a good list, with valid email addresses of people who could plausibly be customers. If you can address well-crafted messages to those people who could use your products or services, direct email is a very effective means of promotion. Other methods of *outbound marketing* such as television advertising are prohibitively expensive for most tech startups, despite rare counter-examples such as the 1984 Superbowl ad introducing the Macintosh.

In other cases, the customer must be *pulled* to a corporate website, or a portion of a site dedicated to a product of service. As suggested by the Guy Kawaski quote at the beginning of this chapter, tech startups are wise to focus on such *inbound marketing*.

A good example is MasterClass, which has exquisitely crafted 2-min trailers showing potential customers what they will see from political masters such as Bill Clinton, culinary experts such as Alice Waters, writing superstars such as Malcolm Gladwell, and the top documentary filmmakers such as Ken Burns.

Social media has aspects of both pull and push. Customers are drawn to your Facebook or Instagram page, or to your YouTube or TikTok videos, if many consumers have widely reposted and shared content that either you produced or that references you. From there on, customers start to engage directly with your content and may choose to follow your brand. Messages and information you post on these platforms will then start to appear in your customers' news feeds. Twitter is particularly effective because it encourages concise nuggets of value for potential customers. Chess.com, Airbnb, Winterlight Labs, and Braze Mobility all send out messages via Twitter.

Yet the fundamental problem is drawing traffic to your website or to your social media posts so that potential customers will notice them, pay attention, and seek more information. Towards that end, vendors of products and services can purchase ads which appear when potential customers input a certain search or phrase or when they view content of a particular kind. Consider Apple as an example. When I am searching for tablets with Google, and likely because my search history includes much exploration of issues related to older adults, one of the ads shown to me is for an iPad tutorial for seniors, which assumes I am an older adult who might want to learn about Apple's tablet offerings.

Key to increasing the likelihood that people will see your message is a process known as *Search Engine Optimization* (SEO). Many people will learn about your products or services by searching the internet. Since they may not know your name, and therefore cannot type it as the target of the search, you need to optimize your website so that you firm' site will come up when a general description of your product or service is requested. There is a similar process known as *App Store Optimization* that applies to a presence on an app store.

Here are a few of the many SEO techniques. Content should include common keywords or keyword phrases to make it more likely that a page will be listed prominently in response to a query. Keywords should also be added to the metadata associated with pages. Pages are more likely to be found if there are more popular pages linking to your pages, so your site should have lots of cross-links connecting its various pages.

A good example of the use of SEO can be seen when one types the phrase "wheelchair collisions" into Google. The first three results are links to material on the Braze Mobility website; the fourth result is a link to a paper written by the firm's founder and CEO, Pooja Viswanathan.

Finally, your message will seem more authoritative if you can get your product or service or firm endorsed directly or indirectly by *influencers* in a particular sector. Influencers are respected because of their knowledge, authority, or fame. They have avid social media followings. Their endorsements, tweets, retweets, and posts are likely to be seen as authoritative and lend credibility to the statements by firms offering products or services. Global influencers such as Gary Kasparov, Bill Clinton, and a scientific authority on dementia could aid the promotional efforts of Chess.com. MasterClass, and Winterlight Labs, respectively.

The best promotion is news that helps establish your authority and credibility and that costs you no money. That's why, for example, SideFX is active in the computer graphics, computer animation, and special effects communities, and why Caseware and its distibutors participate actively in the accounting and auditing professional associations of their countries.

Virality

For inexpensive digital products that can be used or spread on the internet, the goal of promotion is *virality*. Software goes viral when its users make their friends aware of it and encourage them (implicitly or explicitly) to become users. If many of those who learned about it sign up or adopt it, and if they notify their friends in the same way, then the software is said to be *going viral*. LinkedIn, Twitter, Chess.com, Canva, and Wordle all grew substantially through virality. Virality can by stimulated by sharing app-specific content with others, inviting friends to "join your network", using rewards to incenting users to recommend the software to others, or through contests or other forms of socially constructive actions that encourage your users to spread the word to other users.

The simplest technique, in my view, is to embed a token or sign of your app in messages or media that will naturally be sent from a user to other people who could potentially adopt your app. One heavy-handed example is the including of a message such as "This message composed on an iPhone" in messages sent using an iPhone. Social media such as LinkedIn and Twitter transmit tweets and posts and articles created in the medium to people who are not necessarily users, thereby acquainting them with the power of the medium.

Brilliant examples exist in Canva and Wordle. Canva enables people who are not professional designers to create attractive and functional graphics. They will be proud of their creations and likely share them with friends, thus acquainting potential new users with a powerful tool that they too could adopt for free. After someone solves Wordle, there is a Share button which allows the person who played the game to notify others about his or her success with a graphic that displays the solution path without giving away the actual letters of the answer. Telling one's friends about this neat new word game has proved to be an excellent way to spread the word and attract new players.

Achieving virality can pose temptations that are dubious ethically, such as the use of the contacts of an adopter of your software (which can be obtained from many social media companies) to spread the word about your product, with or without the specific permission of the adopter.

Treating Customers Ethically

Most companies assert a commitment to their customers, but few make customer relationships a top priority. For example, Kim Davidson of SideFX believes this is a key differentiator for them: "We stay close to our customers and work to deliver what they value." SideFX also derives enhanced customer knowledge and understanding by working closely with 100 schools and training centers. Pooja Viswanathan of Braze Mobility stresses how much she works on and values her relationships with her customers and says that they are on her phone's speed dial.

Startups must find effective business models. Business models rely upon customers. Customers have needs. If you can address these needs, people will become users; they will pay to meet their needs. Universities and colleges will pay for D2L courseware management systems. Organizations will pay to eliminate long lines with QLess. Pharmaceutical companies will improve their ability to do clinical trials of new drugs that might affect the cognition of seniors by analyzing the speech of trial participants using software from Winterlight Labs. People using wheelchairs will buy smart sensors from Braze Mobility.

Customers will revolt if you do not treat them ethically. You must price your products in a way that is viewed as fair and not predatory, exploiting a dominant or monopoly market position. If you are a B2B business representing a channel to market for your customers, you must not introduce unreasonable restrictions that damage the business of your customers. Both issues will be seen in an analysis of Apple's ethics in Chap. 8. You must be diligent in protecting your customers' data. Facebook, now Meta, has been unscrupulous in selling customer data as a major element of its profitability.

The happiest customers become reference accounts. Their names and experiences are featured on the company's website. They are happy to speak about these experiences with potential customers the firm is wooing. They become, in essence, unpaid members of your sales force. A good example can be seen on the QLess website, featuring as government customers the city of Orlando, FL, and Fairfax County, VA; as education customers Texas A&M University and the University of British Columbia; Office Depot as a retail customer; and Cape Cod Healthcare as a medical customer.

Hence treating customer's ethically is not only good for them but is also a valuable component of a startup's go-to-market strategy.

Strategic Partnerships

Finally, the strategic partnerships of a variety of kinds help startups launch, grow, and solidify their market positions. Some partnerships are based primarily on technology synergies. For example, Apple and Microsoft cooperated at times because Microsoft wanted its software to run well on Apple computers even though the firms competed in desktop, laptop, and tablet computers. Adobe aided its early momentum with a licensing deal with Linotype to gain use of critical typefaces.

Strategic partnerships also yield access to new markets. Caseware has achieved steady international expansion by seeking out entrepreneurial firms with strong technical and accounting expertise and appointing them as partners and distributors in various geographies. D2L partnered with the National Foundation for the Blind to ensure that its technology became accessible to students and faculty with visual impairments.

In summary, an effective business model and a viable go-to-market plan, both supported by creative product design, competitive pricing, an effective presence on the web and in social media, and assertive and aggressive promotion are elements for success. Underpinning this must be ethical and wise financial management, and the successful acquisition of needed financing. I now turn to these latter two topics in Chap. 6.

P. Chess.com

Chess enthusiasts and Stanford Business School students Erik Allebest and Jay Severson had tried various chess sites but felt "at home" in none of them. They therefore launched Chess.com in 2006 by buying the perfect domain name in a bankruptcy sale, putting up a signup sheet, and repurposing it to be a major chess portal, a virtual community, and a chess resource hub.

They created wireframes to clarify their ideas and tried hiring professional programmers to build the first version. Dissatisfied with the results, they wound up doing it themselves (although later with the help of two skilled programmers), foregoing lucrative salaries to build sweat equity in the concept. They relaunched the site in May 2007.

The company has grown over the years by providing a platform of value to chess enthusiasts and new players, and by assembling a variety of resources and new capabilities both by their own internal development and through the acquisition of chess social networking and news sites and chess computational engines.

I am a devoted user of the site. What I most appreciate is the ability to play with friends in other parts of the globe, to be matched instantly against an unknown opponent at a similar skill level, and to see a game I just completed analyzed step by step, showing what I did great, ok, or terribly, and what moves I should have made. With one notable exception (players can drop out of the game without saying so, leaving the opponent sitting for minutes waiting for the next move), the user interface is superb.

Other features, brought in over the years, as the site's content has been continually enriched, include chess news, streamed games by the top masters, tournaments, challenges, chess puzzles, lessons and practice drills at various levels from beginner to expert, leaderboards, blitz chess, 4-person chess, and much more.

Q. Airbnb

Brian Chesky and Joe Gebbia moved to San Francisco in 2007. Noting the scarcity of short-term accommodations for an upcoming design conference, and working together with Nathan Blecharczyk, they decided to use three inflatable airbeds in their living room to make their apartment into a bed-and-breakfast. And in this way they launched Airbedandbreakfast.com in 2008. In their words, 3 guests came as strangers and left as friends. Once they realized that the concept had value more generally than just for conference visitors, and that hosts could and should supply real beds, Airbnb was born. Inspired by the simplicity pioneered by Steve Jobs at Apple, their initial version allowed people to book a home rental with only 3 clicks of a mouse.

Yet it was a struggle. They owed tens of thousands of dollars of credit card debt. They raised $40,000 by personally packing 1000 Obama O and Cap'n McCain's breakfast cereals for the 2008 Democratic National Convention. They commuted from Silicon Valley to New York City weekly for several months to better know their customers who were primarily located there; in other words, they got to know their customers by living with them (see Chap. 3 for how this helped them design their user experience).

Airbnb is a *two-sided business*, in which the internet is used to link sellers with listings and buyers seeking accommodations. EBay was the first, followed by many others including Paypal and Uber.

Eventually, they started to gain traction. They received $20,000 in venture funding from Y Combinator in January 2009. They had 10,000 users and 2500 listings by March and leveraged this to raise $600,000 in April. Later financing rounds were for $7 million, $112 million, $450 million in 2014 (Airbnb was then valued at approximately $10 billion), $1.5 billion, $555 million, $1 billion, eventually going public in 2020.

R. Beyond Meat

After degrees in history and government, the environment, and business, and an early career focusing on clean energy and the environment. Ethan Brown was staggered to read in 2009 that livestock contributed to as much as 51% of greenhouse gas emissions. He therefore founded Beyond Meat in 2009, licensing a meatless protein technology developed by two University of Missouri professors. Brown launched his first product, Chicken-Free Strips, in a Whole Foods location in 2012.

This was followed by expansion nationally in 2013, and the addition of its first plant-based "beef product" in 2014. Investments by wealthy celebrities and athletes and $72 million of venture capital infusions have enabled additions to the product offerings, the opening of new manufacturing plants, and international expansion to over 120,000 retail and food service outlets in 80 countries.

Beyond Meat went public in 2019. Since 2018, the company has been dramatically expanding its research and development capacity. It has also launched strategic partnerships and joint ventures with food giants such as Dunkin' Donuts, McDonald's, Taco Bell, and PepsiCo.

Beyond Meat is a good example of an "ethical technology" startup with a chosen market and product mix explicitly chosen based on the values of the founders.

Finances

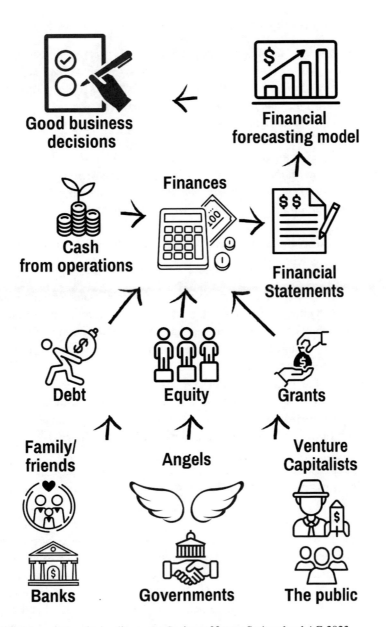

R. Baecker, *Ethical Tech Startup Guide*, Synthesis Lectures on
Professionalism and Career Advancement for Scientists and Engineers,
https://doi.org/10.1007/978-3-031-18780-3_6

Silver's First Law of Venture Capital
$V = P \times S \times E$ where V = valuation
P = the size of the Problem
S = the elegance of the Solution
E = the quality of the Entrepreneurial team

A. David Silver (1985), *Entrepreneurial Megabucks: The 100 Greatest Entrepreneurs of the Last 25 Years*. John Wiley & Sons.

Launched in 2013, the Australian startup Canva supports people in becoming designers of social media posts, brochures, posters, and other graphic creations. Canva grew virally in 2014 from 150,000 to 1,000,000 users, and to 4,000,000 by August 2015. As mentioned in the previous chapter, new and proud designers want to share nifty graphics with friends, thus encouraging rapid growth. The founders were also expert in web marketing. They recruited in April 2014 as Chief Evangelist Guy Kawasaki, a Silicon Valley expert who had been Chief Evangelist for the Macintosh and later a successful investor and VC. They established a very successful Design School to pass on tips and inspiration. All of this drove viral growth, which enabled Canva to raise A$60 million in 2019 and US$200 million in 2021, the latter round valuing the 8-year-old firm at US$40 billion.

This chapter presents the essentials of wise financial management of a startup and of raising the capital needed for growth.

All tech businesses use computers to manage their finances and employ in-house or contract specialists such as accountants to prepare financial statements and tax returns. Yet many do not use financial modeling and forecasting to improve their financial management and their ability to understand what has gone right or wrong with the business. Modeling enables them to realize if the financial results suggest that must pivot their strategy; it illuminates various options for doing so.

The other topic will be the discipline of raising the funds one needs for survival, growth, and expansion. I shall discuss various methods of raising investment capital via equity, debt, and grants from sources including family and friends, angels, venture capitalists, and government. I will also discuss ethical pitfalls that may arise.

Financial Statements and Key Performance Indicators

How does a startup judge its performance? This is done differently in the earliest, *pre-revenue* stage, from the way it is done *post-launch* once the sales of products or services have begun. Judgements can be done qualitatively, with values such as Excellent, Fair, Poor, and Terrible, but they are best made quantitatively. Business performance metrics, including financial metrics, are often known as *key performance indicators*.

Even pre-revenue, a startup's management team has many ways of judging its progress, so long as it has articulated a business plan with goals. Major goals relate to development. If the launch product is to be ready in one year, with progress measured at one-month intervals, then progress at end-of-month (eom) 6 is extraordinary (software is rarely early) if the team has reached the eom 8 target by that time, good if the team as attained eom 6's goal, and terrible the work is 2 months behind what was planned and expected. Other quantitative measures of performance are the numbers of known bugs and the error rates measured during usability testing of early prototypes.

Pre-revenue firms have other ways of measuring their progress. They should have targets for recruiting key personnel, for developing a web site and marketing materials, for connecting with media and influencers, for negotiating partnerships, and for raising capital. All of these should be reviewed monthly to see if progress is according to plan, and if this is not the case, deciding on how to speed things up or to adjust the plan.

Still other critical key indicators relate to the firm's finances. The first one, important pre-revenue as well as post-launch, deals with *cash flow*. How much money is in the bank? How does that compare with what was anticipated and projected? And how quickly is the bank balance changing, in other words, what is the *burn rate*? Not running out of cash is the #1 financial goal of a startup.

Yet there are many other measures of its post-launch success financially. The first one is the rate of *customer acquisition*, especially for mass market apps. Related to that is customer retention or *customer churn*. Significant churn is a sign of real trouble. These measures ultimately drive *revenue growth,* although modern, well-financed startups are often willing to delay revenue in favor of growing a customer base and achieving *market share.* Finally, at some point one must achieve *profitability,* although some visionary startups such as Amazon with huge cash war chests have been willing to wait 5–10 years to attain this goal. Not seeing any revenue for more than 2–3 years or at least break-even for more than 5 years requires a management team with a significant track record of past success and nerves of steel, which also must characterize the investors.

One other measure of a startup's financial viability is its *balance sheet*. The balance sheet computes the *assets* of a company and its *liabilities* (what it owes). The difference is the total *shareholder's equity*, which is a rough measure of what the firm is worth to the shareholders at any point in time. Assets of a software firm are not very meaningful since its major assets are its people. *Profit-and-loss*, cash flow, and the balance sheet are all reported on a regular basis to a startup's management team in *financial statements*.

Honest Financial Reporting

You must be scrupulously honest in your record-keeping, disciplined and comprehensive in measuring financial performance, and brutally realistic in facing the consequences of poor performance. It is critical that you look reality squarely in the eye. You must be open about technology, marketing, and financial problems and risks with your shareholders. The recent scandals surrounding Theranos and FTX and the allegedly inflated metrics of Bolt provide excellent examples of what can happen when you let your beliefs and enthusiasm cloud your judgment. There is a vast difference between founder optimism and securities fraud.

Financial Forecasting Models

The financial statements output by financial management software will tell a firm's principals how they are doing in terms of such measures as cash, customers, revenue, and profit or loss. How do you make sense of these numbers in actionable terms? The answer is to build a *financial forecasting model* in the form of a spreadsheet.

To help entrepreneurs build their own forecasting model, I provide a Microsoft Excel rubric for a prototypical firm TechCorp and a detailed explanation at the web site URL. I show the spreadsheet and provide a terse explanation in an appendix to this book. The model projects the cash flow, profit-and-loss, and balance sheet for TechCorp on a quarterly basis for 3 years. It is quite easy to modify it to project monthly results for 1 year.

The model shows projections for 12 quarters and computes annual totals after each 4 quarters. Years are represented by columns. Rows are organized into sections for Assumptions for Sales, Revenue, and Personnel; followed by the Cash Flow, Profit and Loss calculations, and Balance Sheet calculations. The spreadhsheet has only 60 rows and 22 columns. The assumptions are a mixture of anticipated values of items that are input into cells highlighted with a yellow background and values that are computed by formulas. The formulas are embedded in the cells but are explained in other cells within the spreadsheet and in the appendix. The appendix gives illustrations of how one uses it to pose and answer questions such as those I just discussed. More details may be found on this https://ronbaecker.com/ethical-tech-startup-guide/tech-corp-forecasting-rubric/.

I set up the model so that it works for companies that sell products, those that sell services (including software as a service), those that sell support and training, and those that have diverse sources of revenue. I developed the model when I was teaching in 2011 and have improved it in recent teaching; with my latest startup; and in advising friends starting or contemplating startups including a software venture, a teaching venture, and a media streaming venture. I continue to improve it based on feedback from users.

Many who build financial forecasts think that their purpose is primarily to show venture capitalists to gain investments and bankers to ask for loans. This is false. Their primary role is as a thinking and analysis tool for management. The model is a description of how you believe the firm's finances will develop over time. Building it requires you to estimate many parameters, including how many staff in various categories and compensation levels you will need; how long it will take to build your product and when it will be ready; how quickly and how intensively it will sell, how much you will need to invest in marketing and sales costs to achieve these sales; and what other items such as rent, trade show, travel expenses, and internet marketing costs you will have to pay for.

Such a model represents a static view of your business over time as you envision it at this moment of time. Yet the model becomes a dynamic instrument when you revisit it monthly or quarterly to compare your actual financial performance to what you projected. There will typically be significant or even major discrepancies between the real and the imaginary. Your job, then, as a management team, is to analyze the discrepancies and figure out what was wrong with your assumptions. You then try to improve the model in the hope that it will now better reflect reality. You continue the process every month.

Here is a set of questions that a forecasting model will help with:

How many programmers can I afford to hire given my current financial resources?

Can I afford to give all key employees significant bonuses at the end of this year?

Given my current burn rate, what are the implications of a delay in shipment readiness of the product of 3 months? If this will cause me to run out of cash, how much money do I need so that I do not fail to meet a payroll?

If I believe that $1000 of Google advertising will yield n customers, how much should I invest in advertising?

If I believe that a new salesperson can generate $40,000 in monthly revenue within 6 months, and $100,000 monthly after a year, what are the financial implications of my hiring two more salespeople now?

Should I continue to run my company virtually? What will be the financial implications of getting an office?

Making the model work in this way requires detailed attention to the fine structure of your finances. For example, you need to try to measure how much benefit you garner from certain kinds of marketing and sales expenses, which is not easy to do because changes in monthly customer acquisition may be a result of many confounding factors.

Most startups will not have an accountant or mathematical modeler on their staff in the early stages. It is important that the CEO and the management team do the modeling themselves, as it will force them to think through the interactions among strategy, tactics, venture performance, financial status, and plausible futures. You should build a model as soon as you seriously contemplate a new venture using my spreadsheet or with whatever tools you find congenial. You must then proceed to update and improve it as your understanding increases in the startup planning stage, and also once your venture has launched.

Sources of Funds for Tech Startups

There are many kinds of financing that can help launch a promising startup. I shall begin with government grants, university partnerships, government tax credits, and then discuss loans that result in debt, sweat equity, equity investments, and instruments that are combinations of debt and equity.

The U.S., Canada, the European Union (EU), many of its member states, and many countries around the world have government bodies charged with science, technology, and/or economic development. These agencies typically provide grants to tech startups to develop and launch new products or to enter new markets. Requirements for new product development often include stipulations like those for obtaining patents, i.e., that what is being developed is novel and useful. Granting agencies also typically want to know that the team asking for funds is credible in terms of their ability to commercialize technology once they have developed it and evidence that the technology will work as a solution for the needs of some customers.

Government grants are typically gifts, with no obligation to pay them back, although sometimes there are repayment or royalty provisions. Other grants are available from private foundations and are pure gifts. Gifts are free money, so they are highly desirable, but they are highly competitive. The government grant decision process is also typically very slow, sometimes taking more than a year, so they are most suitable for funding research and long-term development that is relevant to the next generation of products.

Another "free" resource is a university partnership. Again, such an arrangement is best used to fund research, as that fits a university's mission. If you want to highly motivate a professor to enter a collaboration, consider a sweetener such as a promise of later consulting or a modest equity stake, but beware of breaches of ethical conduct. Another mutually beneficial outcome of a university-startup partnership is that students have a direct path to interesting jobs and that the startup has access to a pool of good young talent. Many of our example startups have benefited greatly from university partnerships, for example, Microsoft, Adobe, SideFX, Casewate, D2L, Winterlight Labs, Blue Rock Therapeutics, and Braze Mobility. Others, especially firms in Silicon Valley such as Apple and LinkedIn, or in locales with quality universities and lots of tech startups such as Seattle, New York City, Toronto, Boston, or Cambridge UK, benefit significantly by being close to sources of new, young, well-educated talent.

Some governments also support startups needing to do research by offering tax credits which offset some of the costs of the research. In Canada, for example, firms doing what the government considers R&D can get back on the order of $1 for every $3 of expenditures if the work qualifies for the Scientific Research and Experimental Development (SR&ED) tax credit. As with government grants, there are complex rules to qualify, so it is often desirable to hire a consultant knowledgeable with the rules and the process.

Debt and Equity Financing Methods and Stages

In a way, the simplest way to get needed funds is to borrow them, in other words, to assume debt. Terms must be negotiated, including the interest rate and when the loan must be repaid. This often works well for "family and friends" financing. Banks, however, are loath and mostly unwilling to lend money to startup tech firms, and will only do it with significant security, typically a lien on one or more of the homes of partners. In other words, if you do not repay the loan, you lose your home. Independent of the interest rate, such money is not cheap.

Another way of building your product without needing to raise a lot of money is known as *sweat equity*. Sweat equity is the value accrued by company founders working evenings and weekends without getting paid. These efforts can continue for months or years until the venture has sufficient capital to pay founders a salary. We have seen examples of the power of sweat equity with startups such as Desire2Learn, Airbnb, Chess.com, and Winterlight Labs.

A startup's initial equity, or ownership, is divided among the founding partners. They receive the potential value of this ownership as a reward for their ideas, their initiative in forming the venture, their creativity, and their unpaid work. There are two ways of dividing up the pie. One is that each founder gets an equal share. This has the advantage of securing significant commitment for everyone, without needing to decide whose work and skills are worth a greater share. The other way is to divide the equity in proportion of each person's worth. Both run the risks that some founders will think, at one point of time or another, that they have not been treated appropriately and fairly. There is no magic remedy to this danger other than open and honest communication among the founders, with the goal of finding a solution that is at least reasonably acceptable to all partners.

To enable growth, almost all startups raise external funding, although Wordle is an example of a new venture creation done simply with sweat equity. Typical initial funding can amount to tens of thousands to tens of millions of dollars. In launching Nuula, Mark Ruddock raised US$20 million in an equity infusion by a consortium of venture capitalists and a US$100 million line of credit, which gives him access to that amount of debt. Blue Rock Therapeutics, requiring capital to fund perhaps a decade of research and clinical trials, raised US$225 million in Series A venture funding from Versant Ventures and the pharmaceutical giant Bayer AG.

Every equity infusion is preceded by a negotiation between the startup and its potential investor about the *pre-money valuation* of the firm. Let's imagine that you and a partner have worked evenings and weekends over a year developing a first version of a product that seems to have a market without killer competition. The $ value of the time you have invested—your sweat equity—may be $150,000, but your *founders' equity*— the sum of your sweat equity, the value of your idea and your work up to that point in time, and the value of your expertise, background, and track record—may be far greater, for example, $900,000. Hence the investor advancing $100,000 would garner a 10% ownership share.

Successive investment rounds are best staged with reference to the achievement of corporate *milestones*. Returning to this example, the founders would logically seek an initial set of customers for their product. They would use the $100,000 to achieve this goal. After successfully doing so, they might assert a valuation of $1,800,000 and seek $200,000 to fund market expansion, so in this case a 10% share would be given up in exchange for a larger sum. Ideally, successive rounds are done at higher and higher valuations. If these investments enable significant growth of the firm, then the founders' smaller ownership share is ultimately worth more, so everybody is happy.

Finally, and this also is an advanced topic, there are *combination instruments* in which funds are advanced to a startup that are both debt and equity. Here is a simple example of *convertible debt*. Your rich aunt agrees to loan you $150,000 to pay for sales and marketing expenses to launch the product you have developed over some time with the sweat equity of evening and weekend work. She agrees to charge you no interest for a year, but then wants 10% interest per year. You agree that the firm's potential is so great that pre-launch it is worth $850,000. There is a conversion privilege, which allows the loan to be forgiven if the amount is turned into a 15% ownership stake in the startup. The actual situation is more complex than I have space for here, as there must be agreement on who has the right to insist on repayment of the loan or conversion into equity, and when and under what conditions this can be done.

Most startups that achieve success and growth need cash at numerous times; hence there are *financing stages*. Although every case is different, here is a typical pattern.

The first is often a *family and friends'* stage. These are typically in the amounts of $10,000 to $250,000, and are funds advanced by people close to the founders as debt or equity or both. It is vital that the founders stress that the investment is speculative and risky. Another critical issue to talk about is when the amount might be repaid or when the shares might become liquid through the company's acquisition of going public. I personally was naive in investing what for me were significant amounts at the age of 65 in the startups of second cousins and later being frustrated at how long I would have to wait to at least get my money back (if at all) or to see any substantial return. You want to be very clear to discuss the risks when discussing loans or investments from family.

The next stage is one or more *angel investment rounds.* These are typically in amounts ranging from $25,000 to $750,000 and are made by individuals known as angel investors. Typical angels will have made significant amounts of money from one or more past startups. They seek with their involvement to make more money, but the word "angel" is used because they also want to give something back, to convey what they have learned to new young entrepreneurs and help them by providing advice. Angels advancing significant funds will likely want board seats; their expertise will aid the deliberations of the board.

A financing option not widely used is to obtain a *strategic investment* from a firm that is in your market area but is not competitive. A great example is the purchase of a minority share in SideFX by Epic Games, a much larger developer of computer games. The investment was sought by SideFX to achieve greater financial stability and to allow one of the founders and other shareholders to liquidate their shares and see a financial return. Strategic investments may sometimes be obtained in order to finance the development of a product required by a larger firm that does not have the capability of building the product themselves. Strategic investments also serve to open new market and partnership opportunities for startups.

Another and sometimes final stage is one or more *venture capital* rounds, typically known as Seed, Series A, Series B, etc. Each stage is usually for a larger amount of money. The amounts invested typically range from $1,00,000 to hundreds of millions of dollars. An initial investment at the low end of this range is known as *seed funding*.

Venture capitalists (VCs) are individuals who theoretically are expert in business and in certain industrial domains, such as information technology, the internet, biotechnology, or new forms of energy. Their firms raise capital in funds of typically $100 million to $1 billion from much larger institutional investors such as pension funds. VCs then invest money in early stages of growth companies with the goal of exiting (this concept is discussed below) in 5–7 years, having then made 10–100 times what they invested. It may seem that VC expectations are high, but experience has shown that despite their expertise at least 8 out of 10 of the firms that enter a VC's investment portfolio will either go bankrupt relatively quickly or join the legion of the *living dead*, companies that hang on for dear life seemingly forever.

With both angels and venture capitalists it is essential to remember that once they have given you substantial amounts of money that they are now your partners. This means that you want to be certain of their integrity, ethics, and good judgment before you take their money, and that you need to consult them before taking critical strategic decisions. They will typically want board seats. You will also want to provide them with an exit strategy of going public, a merger, or an acquisition, to be discussed below.

Values, Ethics, and Your Funding Partners

Most recent startups that have managed to grow rapidly and achieve dominant market positions have done so with the help of investments from angel investors and venture capitalists. For example, in part because founder Reid Hoffman had good connections to venture capitalists, LinkedIn was able to obtain a Series A venture capital syndicated investment in late 2003 less than a year after its founding. The firm later received a $53 million syndicated VC infusion in June 2008; obtained one more investment round in 2010; and went public in May 2011. Blue Rock Therapeutics received a huge venture investment from a venture firm and Bayer AG, a major international manufacturer of pharmaceuticals. Airbnb's investment stages are described in Case Q; MasterClass's investment rounds are presented in Case T.

It is important to understand what to look for and what to avoid in choosing a VC investor. Desirable attributes are good industry knowledge, hands-on experience in running startups, deep pockets (so they can fund future rounds), collaborations with other VCs (so they can *syndicate* rounds among multiple parties), extensive contacts within the industry (so they can help you in team building), honesty, openness, humanity, past success, and happy investees.

You must do reference checking and speak with several firms that the VC has invested in. You want to understand their values, their ethics, and their policies and procedures. Particularly important is to understand how the VCs behaved when things were bad, when the firm was in trouble. Were the VCs patient and constructive? Did they have staying power? Be sure to ask about the VC partner who will lead your investment, and not just the firm. Things to avoid are the lack of these attributes, especially the personal qualities and the direct startup experience. You must enter the partnership with full realization of the fact that they will give little weight to your role as a founder and try to change the management team (and usually succeed) if they believe you are not delivering optimal financial performance.

It is important that you be comfortable with the ethical stance of potential funders before taking their money. You do not want investment from individuals or firms who will encourage actions you would view as evil. You do not want as your partner a VC that has a record of financial manipulation that devalues the stake of early investors. (I know of one such case in detail, but I cannot discuss it here.) You must understand and agree with the growth, profitability, and liquidity expectations of potential funders, as "growth at any cost" may profoundly change the nature of your business and transform your creation into one you would neither recognize nor respect.

Founders should also keep in mind that despite their own ethical stance, they may not always remain in control of the business they are creating. The need for growth and the need for an exit may put control in the hands of third parties—professional managers, financial managers, or the board—whose values and ethics do not align with yours. Choose your partners carefully, as they may ultimately be responsible for carrying your vision—and your ethics—forward. Build ethics into your company from the ground up. Companies with brittle ethics depend on founders continually making the "right" decisions. Companies with strong ethics build ethics into the product or service itself and build a culture and values that encourage "good" decisions to be made at all levels of the firm, by an entire team aligned to doing the "right" thing.

The investment deal with a venture capitalist is called a *term sheet* and is often only one or two pages long. The contractual investment agreement may be hundreds of pages long full of clauses such as rights of first refusal, anti-dilution clauses, and drag along provisions, which I do not have space to discuss. You need a competent corporate lawyer (not your family's attorney) who has deep experience with startup financing to conclude the deal.

Growth Strategies

A tech startup needs to grow to survive. *Lifestyle businesses*, such as local restaurants and hair salons, can provide their proprietors stable incomes even if they do not expand. They can compete with large chains operating in their industries because of personal service, individuality in style and substance, and customer loyalty to a business that is owned and run locally. In fact, they may dilute their appeal if their individuality and customer knowledge and service is subverted to goals of expansion.

Yet growth strategies are essential for tech startups because technology is a rapacious industry with ambitious entrepreneurs such as yourselves and large behemoths such as Amazon, Facebook, and Google eager to copy your ideas and roll over you using their superior might and vast supplies of capital (the cash they have in the war chests). Alternatively, they seek to scare you to death and then buy your firm at the cheapest price possible. An effective growth strategy consists of stages of innovation, expansion, and consolidation; likely followed by an infusion of capital to help you get to the next level; followed by more innovation, expansion, and consolidation. The stages overlap. It is important to have these stages in mind early on, as you risk your momentum stalling if you embark on one stage without a financing plan to get to the next stage.

Tech wizards who did not want to engage in the business of such growth and the attendant stresses that accompany it are well advised to avoid accepting external capital and establish a consulting or contract programming lifestyle business, or simply to work for somebody else.

Exit Strategies

When I started my first software firm, I never thought of the end game. I wasn't thinking of getting rich, I just wanted to create things of value and get people using them. I paid no attention to what might happen in 5 or 10 or 20 years. Did I imagine we would grow forever? I gave it no thought. I had no *exit strategy* in mind. This was a serious mistake.

There are five exit strategies. The purpose of an exit is to allow entrepreneurs to move on to the next stages of their lives, including moving to another job, changing your profession, creating another startup, or retiring. Ideally, entrepreneurs exit with some cash, which is particularly useful for the last three options.

The first option is bankruptcy. You get no cash, and there are dismal legal repercussions of having to file for bankruptcy, which will include damage to your credit rating and may include loss of your home.

The second option, as I have mentioned, is what VCs call the *living dead*. This means that you do not go bankrupt, but your startup is hanging by its fingernails, often being uncertain if it can meet its monthly payroll. This is a dreadful way to exist for more than a few months. It may be better to declare bankruptcy.

The exit strategy that is not an exit is to continue forever. If your products and services are in a domain that is not highly competitive, you may have an acceptable lifestyle business, giving you satisfying work and a good salary. If the area is highly competitive, you may need to grow by raising more capital. Your investors will likely then pressure you towards one of the two remaining strategies, going public or being acquired.

One financially desirable exit is to *go public*. The process is to register with a stock exchange such as the New York Stock Exchange or NASDAQ and offer shares to the public at a price that is attractive to both the founders, any venture capital partners who have invested in you previously, and the brokers and analysts who advise people on what stocks to buy and when. One reason that many tech firms are reluctant to go public is that it necessitates a huge amount of regular reporting (every 3 months) to satisfy regulatory agencies and to allow shareholders, prospective shareholders, knowledgeable investors, and potential investors to decide whether to buy. sell, or do nothing. There is also legal exposure for principals of the company to not disclose confidential information to people who could benefit from this information. Antibe Therapuetics went public relatively early in its existence because of a relative lack of biotech-focused Canadian venture capital at that time.

Going public allows your stock to become liquid, i.e., you can sell as least some shares and turn them into cash. There are limits to liquidity, as principals who are major shareholders often have restrictions on when they can sell part of the holdings and how much they can sell. A consequence of going public is that your corporate goals, strategies, and tactics will change. Whereas smart private investors such as angels and VCs may have the patience to see you forego short-term gains in favor of longer-range planning and investing, the public markets and the analysts who write about them are ruthless in demanding continual positive financial results.

Another way to gain liquidity is to merge with or be acquired by a larger firm that has a more secure market position. For founders whose dream is to make their products universal, this can be very attractive, although the acquiring firm will often consolidate operations resulting in loss of some jobs. Liquidity may be immediate, if the acquisition offer includes significant cash, but more often it totally or primarily consists of shares in the acquiring company. In this case, the founders can cash out only insofar as the stock of the acquirers is publicly traded.

Seeking a well-resourced and stable firm to buy you can also be done for reasons of stability and growth. A good example is LinkedIn, which protected itself against the power of Facebook by seeking and succeeding in being acquired by Microsoft. Another example is Blue Rock Therapeutics, which was acquired in 2019 by Bayer AG, which previously had been a significant early-stage investor.

In summary, financial management must be active and not reactive. Wise startup teams will build forecasting models which encapsulate their understanding of the numbers underlying their businesses, will compare as often as monthly their performance with the models' predictions, and will use the discrepancies to advance their understanding of the businesses and to inform strategic and tactical decisions.

Startups other than lifestyle firms will need infusions of cash to develop innovative products, to fund sales and marketing, to withstand dips in their success, and to fuel growth. There are many ways to this, including government grants, sweat equity, help from family and friends, angel investments, venture capital investments, and going public, all of which have their advantages and disadvantages. It is important to understand this range of options, and to employ various methods at appropriate times as you grow.

Most important is to remember that all investors are now your business partners, and that you then must consult them and consider their needs even if you still formally control the business. This is one of many reasons that investors should be chosen based on their integrity and ethics as well as the depth of their pockets.

S. Canva

Canva is a user experience company. Created in response to the increasing functionality, complexity, and cost of tools from firms such as Adobe, Canva was launched in 2013. Its goal was to allow almost anyone to become a designer of social media posts, brochures, posters, and other creations involving the structured display of text and image.

To do this, it had to make the initial experience so successful and so magical that it could overcome the widespread feeling among so many non-designers that they cannot do design. They had to communicate—"Yes I can!".

Canva uses a freemium business model. This means that anyone can use the software with most of its functionality for free forever. Advanced functionality is available for a monthly cost at two levels, one for individuals or groups of up to five people, and one for enterprises.

Canva grew virally in 2014 from 150,000 to 1,000,000 users, and to 4,000,000 by August 2015. There were several reasons for this success. As previously discussed, novice designers are proud and want to show nifty graphics to their friends, thus encouraging viral growth. The founders were experts in web marketing. They recruited in April 2014 as Chief Evangelist Guy Kawasaki, a Silicon Valley expert who had been Chief Evangelist for the Macintosh and later a successful investor and VC. Finally, they established a very successful Design School to pass on tips and inspiration.

Canva raised A$60 million in 2019 and US$200 million in 2021, the latter round valuing the firm at the extraordinary figure of US$40 billion, an amazing valuation for a firm so far away from Silicon Valley.

T. MasterClass

David Rogier worked briefly for a venture capital firm after graduating from Stanford, then did an unsuccessful startup developing a device for people with allergies. Contemplating a new venture in online adult education, which he was passionate about improving, he then paid 12 respondents on Craigslist to tell him about their user experience with adult education. Together with a technical co-founder, they then focused on the "crazy idea" of getting the best people in the world as their instructors.

Rogier's original target was a trio of James Patterson, Serena Williams, and Dustin Hoffman. It was a hard sell, but 3 years of sweat equity, resourcefulness, an unwillingness to take no for an answer, and with the help of $1.5 million of venture financing, MasterClass launched in 2015 with 3 classes available to view at $90 each.

Instructors now include such luminaries as Bill Clinton, Frank Gehry, Wayne Gretzky, Garry Kasparov, Steve Martin, and Penn and Teller. MasterClass's revenue model is a subscription for $180 per year, which was reduced to $1 during the early stages of the pandemic. 200,000 college students signed up in one day. It now has 130 courses and 1.5 million subscribers.

The firm has been remarkably successful in attracting investors. It has had 2 angel rounds and venture capital rounds Series A through F, for total invested capital of $240 million, giving it a $2.25 billion valuation as of May 2021.

U. Winterlight Labs

Dr. Frank Rudzicz is an artificial intelligence and language processing expert working primarily as a hospital research scientist. He also supervises graduate students in the University of Toronto Department of Computer Science. He has long had an interest in methods for diagnosing and monitoring cognitive decline, which often results in some form of dementia such as Alzheimer's disease.

Standard methods involve administering short tests to individuals concerned about their cognition, the most common of which is the Mini-Mental Status Exam. If results on this test are indicative of a problem, then a full 3-h cognitive assessment including a battery of tests is done by a cognitive neurology expert. The word "battery" is a fair assessment of the feelings of people who need undergo this testing. If there is a need for reassessment a year later, the entire process is repeated.

Inspired by previous research indicating that cognitive decline could be detected in writing, for example the novels of Agatha Christie over time, Frank began research in 2013 on using attributes of short snippets of speech as biomarkers to signal cognitive decline. Some of the work was done by two of his graduate students. Results were encouraging, so the three of them formed Winterlight Labs in 2015. They were soon joined by Liam Kaufman, whose background was in psychology, medical science, computer science, and entrepreneurship, with one successful software startup as developer and CEO.

Winterlight's tools enable scientists and clinicians to track, screen for, and predict cognitive decline and other neurological changes efficiently, and are currently being used mainly by pharmaceutical companies in clinical trials of new medications for staving off and treating dementia.

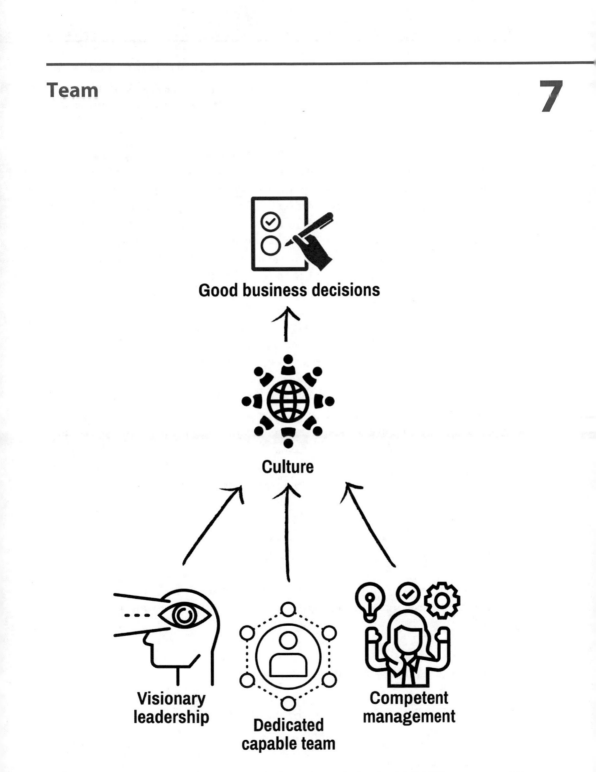

Good business decisions

Culture

Visionary leadership

Dedicated capable team

Competent management

© The Author(s), under exclusive license to Springer Nature Switzerland AG 2023
R. Baecker, *Ethical Tech Startup Guide*, Synthesis Lectures on
Professionalism and Career Advancement for Scientists and Engineers,
https://doi.org/10.1007/978-3-031-18780-3_7

> *"Culture is what people do when no one is looking."*
>
> Gerard Seijts, Executive Director, Ian O Ihnatowycz
> Institute for Leadership, Ivey School of Business,
> Western University, Canada

Toronto software developer SideFX's speciality of procedural technology has become more and more important as the complexity of computer-animated scenes has grown to include crowds of huge numbers of objects such as people, animals, birds, buildings, roads, trees, clouds, and mountains. Hence, for 35 years, SideFX has been an excellent example of a small startup controlling its own destiny with no venture funding, succeeding by focusing on excellence in a specific market niche, under the steady hand of leader Kim Davidson. But great leaders are often challenged by serious crises, as happened to SideFX in the early 2000s, when it survived a financial crisis by trimming staff, holding raises, and introducing variable bonuses, but chiefly because Kim had architected a culture of humanity and shared purpose.

This chapter explores the nature and importance of visionary leadership, skillful management, and a talented, energetic, committed, and well-compensated team. I shall explain that leadership and management are very different. These skills are often not found in the same individuals.

Leaders articulate the vison and motivate the team to excel in pursuit of the vision. They set and mold the corporate *culture*. Startups also require a talented management team, backed up by employees who possess both extraordinary skills in their disciplines and huge drive and commitment to the demands and stresses of a startup. Among the most critical issues that management sometimes must face is whether and when they should pivot the strategic direction of a venture, and how to behave ethically in times of corporate stress.

Leadership

A lovely description of leadership appears in a Anna Mar's 2013 "7 Definitions of Leadership". It says that leadership seeks to "get a group moving in a common direction".

1. Leaders Motivate
   A leader is a dealer in hope.—Napoleon Bonaparte
2. Leaders Make Decisions
   Leadership is doing the right things.—Peter Drucker
3. Leaders Coach
   Good leadership consists of showing average people how to do the work of superior people.—John D. Rockefeller
4. Leaders Are Confident

A man who wants to lead the orchestra must turn his back on the crowd.—Max Lucado

5. Leaders Influence

   A leader is best when people barely know he exists, when his work is done, his aim fulfilled, they will say: we did it ourselves.—Lao Tzu

6. Leaders Innovate

   Innovation distinguishes between a leader and a follower.—Steve Jobs

7. Leaders Get Results

   Leadership is defined by results not attributes.—Peter Drucker.

How do you get a group moving tirelessly and with a vision of a difficult but plausibly attainable goal? You demonstrate your expertise and your ability to innovate. You exhibit confidence. You make decisions sufficiently promptly, even in the face of uncertainty, avoiding the paralysis of indecision. (This was hard for me with my first startup; as an academic turned entrepreneur, I often sought more information and the illusion of certainty before making decisions.) You get results, building confidence in the viability and promise of the firm to get even better results. You are a great coach, teaching and inspiring members of the management team and everyone in the company.

Leaders model ethical behavior and inspire others to do the same. A good counter-example is Travis Kalanick of Uber, who created a culture of unethical actions, seeking to destroy the taxi industry independent of the consequences (such as the suicides of 6 New York city taxi and limo drivers within 1 year), indulging in unreasonable corporate behavior (such as spying on and sabotaging competitors), fighting attempts at reasonable oversight by government, and disobeying laws. While he created a valuable company, his successor has now spent years trying to undo the damage done internally to the morale of employees and externally to the public.

How do you set the direction? Here is one place where it helps to have a management team (to be discussed below) that brings different perspectives to bear on the decision. What do customers want? What can you create for them that will be innovative and compelling? What resources are needed and how long will it take? There will never **the** correct answer; at some point, you must make the best decision you can make and move full speed ahead.

Leaders are not always gentle. They can be abrasive and even demeaning. Bill Gates and Steve Jobs are two good examples of this, as their desire for perfection and high standards often resulted in brutal attacks on people in meetings. But employees put up with them because of their vision (a shared vision), their drive, their clear and outstanding expertise, and their track record of success.

Caseware's CEO and COO jointly delivered greater and greater product innovation and market success worldwide. The CEO provided the technical vision and leadership, the COO a steady hand on operations and business affairs. Both also had an abiding and resourceful commitment to making Caseware an excellent place to work, which resulted in very low employee turnover.

SideFX's Kim Davidson is a great leader. He writes about motivation, vision, and focus; and about overcoming his greatest challenge.

SideFX was started by myself and a co-founder in 1987 ... with nothing but an idea and a passion. We had no prior business experience. We were two people who loved making animated art with computers.

Over thirty years later ... I and others keep the entrepreneurial spirit alive and well at SideFX. Every year brings new challenges to our company - new competition, hardware, and regulations. Ensuring that entrepreneurial spirit is propagated throughout the company is critical to meeting these challenges.

We compete against much larger competitors by focusing on specific production workflows where we build and maintain a significant competitive advantage. We do this by continually and rapidly adding powerful innovative features onto a mature and stable platform honed over three decades.

We overcame a financial challenge in the early 2000s — having been close to bankruptcy — by reducing staff, holding raises, and shifting to variable bonuses. Since then, we have focused on building financial resilience, and have been in a consistently strong and growing financial position for more than 15 years.

Richard Hamel, previously Vice President of Sales and Marketing of SideFX (for 24 years), describes Kim in this way:

[Kim always knew] what mattered most to him - the longevity of the company, his staff, and serving his customers with integrity. ... he has always been clear about which few to focus on and excel at. His humbleness, intelligence, easy-going calm manner and his genuine love of animation and visual effects have been instrumental in creating an enviable culture at SideFX, full of loyal staff pleased to serve under his leadership.

Cristin Barghiel, Vice President of Research and Development of SideFX (for 29 years) adds:

Kim is a leader who stands firmly at the head of his company. Steadfast, he is the calm person at the table with a clear long-term vision for his company both in terms of VFX technology and corporate culture. It was Kim's early imprint of equal, indelible focus on art, science, and culture — uniquely approachable, casual, almost familial — that made this company what it is. Kim's masterful ability to nurture this fragile mix into a defining corporate attitude toward client interaction and inward growth are testament not only to his vision but resilience — some might say brilliant stubbornness — in the face of internal and external change.

Few companies have been able to retain their identity with passing decades as well as SideFX: even as the company grew, it managed to retain the same genuine connectedness within its ranks and in the marketplace. One would have difficulty naming SideFX's defining quality — fast paced innovation in the face of much larger competition, eminently close to the needs of the community, nurturing corporate culture — precisely because to Kim all three have been equally crucial from day one.

Culture

Culture is key to startup success. Here are the most essential factors that result in a good startup culture.

The leader should be a "mensch", which is a Yiddish word meaning "a good person, a person of integrity and honour". Menschen garner respect, essential when one asks a team for extraordinary commitment and performance.

A good culture requires also that the leaders have respect for others, for every member of the team. It is essential that everyone be treated well and fairly. Values play a key role in the culture of successful startups.

There also should be an inspirational vision, and a standard of excellence that is known and celebrated. Caseware's CEO has noted that its vibrant culture with almost no turnover was due to an environment encouraging the solution of tough problems and seeking a dramatic improvement in the ways auditors and accountants do their work.

Culture is critical to scalability, because startups that experience viral growth are hiring so fast that everything cannot be explicitly communicated. New hires must be guided by what they see and sense around them and perform well when nobody is looking.

Kim Davidson writes about the SideFX culture:

> *Five years ago, I worked with* the staff *over many months to co-create our* purpose, vision and values. *Along with these, the staff concretized our values with exemplars and a description of our future self. I provided little direction and, as a result, the staff is highly invested in the results.*
>
> *My philosophy is that I work for* all SideFX stakeholders. *This* includes shareholders, employees, customers, *suppliers* and partners. ... *I look for consensus in decision making and will go with the majority on a decision even when it is not aligned with my choice. I believe in collective intelligence.*

Richard Hamel writes about the SideFX culture:

> *The low turnover for a company that is full of bright and talented people, many of whom could make more money elsewhere, speaks to its appeal. Kim has undeniably nurtured its culture. SideFX is a big, evolving family for Kim. He wants* the environment *to be stimulating, innovative, respectful, fun, non-corporate, and collaborative. He'll sit beside a new interning artist and learn about them and their aspirations. It might be a week or two before the intern realizes he is the CEO.*
>
> *... the culture has also been shaped [because] Kim is not pursuing personal wealth. ... That gives everyone a sense of his unwavering love for the company, its people and the customers. He believes in the company doing the right things that serve the long term.*

Pivoting

A critical issue for leadership is *pivoting* the strategic direction of a company. This typically happens when the then current strategy is not achieving desired goals with respect to metrics such as market share, revenue growth, or profitability. If the firm is truly floundering, a successful pivot is called a *turnaround*. Sometimes the attempt of achieve a turnaround is triggered by the board and the controlling shareholders through a change in management. Here are some instances in the life of our case study companies.

Microsoft: The company's IPO in 1986 made 12,000 employees millionaires. The brilliant co-founder, visionary, and long-time architect of its success, Bill Gates, handed over the CEO reins to his college buddy Steve Ballmer in January of 2000. In April, a judgment was handed down in United States v. Microsoft Corp., calling the firm an "abusive monopoly". Although the parties settled the dispute in 2004, the European Union also soon brought successful antitrust action against Microsoft. There followed a decade of slow revenue growth, an inability to move vigorously in the new internet and mobile markets, over a decade with a stagnant stock price, its first ever quarterly loss in July 2012, its worst stock price day ever in July 2013, and a layoff of 14% of its workforce in July 2014. Ballmer was replaced in February 2014 by Satya Nadella, who had been head of its Cloud Division.

Nadella can be credited with a turnaround in Microsoft's fortunes. Despite the 2014 purchase of Nokia Devices and Services pushed through by Ballmer, he abandoned the attempt to compete in the mobile device space. He has grown the firm's presence in cloud computing as a strong #2 behind Amazon Web Services, led the company in consolidating its strong position in large corporate computing, emphasized agile programming, and added new vigor in gaming. Yet in my view his major success has been in making collaboration and empathy central to the corporate culture, and in emphasizing the firm's ethical responsibilities with respect to domains such as artificial intelligence and cyberwarfare.

At a time when Big Tech typified by Amazon, Google, Facebook, and even Apple are under serious attack for moral lapses and indifference, Microsoft under Nadella has emanated a sense that it is a firm with a soul and with conscience. All of this has been rewarded with a huge increase in stock price and corporate capitalization.

LinkedIn: Its founding CEO Reid Hoffman is widely respected in Silicon Valley as a visionary and savvy early-stage investor. One aspect of Microsoft's 2016 acquisition of LinkedIn for an astonishing $26 billion was the appointment of Hoffman to the Microsoft Board. The deal was a courageous one for both companies. For Microsoft it was a high price, but one that seemed to make sense, in that it added a social media piece to its corporate computing arsenal. For LinkedIn, they risked being absorbed with resulting invisibility. Yet they received great returns for their shareholders, and access to resources and deep pockets to battle the Facebook juggernaut. They were promised independence from their new owner (which seems to have been respected), and have thrived in terms of revenue growth.

Antibe Therapeutics: Antibe is a drug discovery company seeking medications that target pain and inflammation. Its non-steroidal formula, like others of its kind, can cause a rise in toxic liver enzymes. Antibe was originally going after the chronic pain market. It did not expect serious side effects at lower dose levels, yet these appeared in their clinical trials to a degree that was troublesome. CEO Dan Legault therefore pivoted Antibe in 2021 to go after the smaller acute market, targeting post-operative pain, where the side effects of short-term usage would not be significant. Work continues to find a path to treating chronic pain. Clinical results show good pain reduction in chronic pain without compromising the gastrointestinal tract.

Nuula is a startup, having been created in early 2021, even though it was formed out on a pre-existing firm. As described in case study X, its initial mission was to deliver both a super-app providing succinct and accessible mission-critical information and a Line of Credit for emergency short-term funding. Courageously, CEO Mark Ruddock presented to the board early in 2022, one year after the launch, a proposal that the firm was too thin in focus, management attention, and capital to continue to do both functions, and that it should henceforth outsource the Line of Credit. This huge pivot was accepted by the board.

Management Teams

Most startups have several founders, but only one leader. Microsoft was founded by Bill Gates and Paul Allen, but the technical savvy, business acumen, and leadership of Gates drove the company's success for 33 years. Apple could not have become a $3 billion tech giant without the engineering talent of co-founder Steve Wozniak, but it was the drive, design sense, marketing ability, and leadership of Steve Jobs that propelled Apple to the top, despite years during when it seemed that Apple was a good idea that could not compete with the juggernaut of the many firms manufacturing IBM PC-compatible personal computers.

Yet leaders cannot run successful companies on their own. When I founded my first startup HCR and ran it for its first 8 years, I could do this in part because up to about 30 people could fit into a conference room. But when we grew to more than this number in 1981, I needed a management team who shared in the making of key strategic decisions, the execution of tactics, and the supervision of other key managers.

Management teams often grow in stages. It can begin with one founder, such as Alex Backer of QLess and Drisit, but the founder must soon enlarge the team. An excellent pattern, especially for academics, is to find someone with a good combination of domain knowledge and business savvy, as did researchers Dr. Frank Rudzicz when he found Liam Kaufman to run Winterlight Labs (Liam had already created one successful startup), or when Dr. John L. Wallace recruited his friend Dan Legault, an experienced business executive, to take charge of Antibe Therapeutics. Typical founding pairs are two technical co-founders, such as Kim Davidson and Greg Hermanovic of SideFX, but it helps if one of them has some of the skills required for sales and marketing. Another classic founding duo has one person expert in technology and another in business, and especially in sales and marketing, which was the case with Steves Wozniak and Jobs of Apple. Entrepreneurial romantic couples can often succeed as a team, especially if their skills are complementary.

By the time the company is mature, it is reasonable for it to have a CEO, a VP of Technology or Engineering or Software, a VP of Sales and Marketing (or two individuals, one with each portfolio), and a VP of Finance. (At the start-up stage, the finance portfolio is the easiest to outsource to a consultant.) Another important addition to the team is a COO (Chief Operating Officer), to offload some of the CEO's administrative duties and allow him or her to focus more on strategy, high-level sales, and setting the technical direction. Visionary firms will include a VP or Director of Products or of Design to ensure that user experience and customer experience are considered at the highest level of decision-making (At Apple, CEO Steve Jobs pl;ayed a key role here, partnering with Jony Ive, their long-time head of design.)

The management team has many roles. It discusses and determines corporate strategy, as well as needed course corrections as they arise. It reviews, likely on a weekly basis, the progress that each department has made towards meeting its objectives. It weighs the advantages and disadvantages of sales and marketing approaches. It makes decisions that affect the company's financial position, such as how quickly to hire people and what to do if layoffs seem necessary. It guides the development of product enhancements and new technology directions.

Each member of the management team needs to be expert in his or her domain. Yet, because of the breadth of decisions a team must take, it is helpful if each member is or can become somewhat of a generalist so that discussions can be animated with multiple ideas and points of view.

The management team also must inspire, direct, and mentor members of their departments. As the company grows, the workload increases. Key to success is not micro-managing, but rather *delegating responsibility and authority*. In other words, you must make crystal clear the deliverables and deadlines for people reporting to you, and then give them the authority and the resources to do what is necessary to accomplish their missions. You need to review on a regular basis, preferably weekly, how well they are doing, and guide them with firm and honest appraisals of successes and areas that need improvement. You must ensure that they have the resources (most importantly, personnel) to do the job, and if necessary, remove them from the positions if they are not doing the job. As a startup grows, managers at each level whose staff are getting beyond a reasonable span of control, typically 5–20 people, will need to appoint members of their teams to be managers of small groups, a process that continues as a company grows.

Cristin Barghiel writes about Kim Davidson of SideFX:

> *Kim's management style is relaxed. He is as far removed from micro-managing as one could. He is understanding of individual and context, accepts and invites learning by mistake, and always, always shows caring and compassion. He is there to steer if need be, and has an uncanny ability to know when that is the case, and otherwise leads by helping you grow your own wings.*
>
> *Kim is there to help, to advise, even to humour. He is no more hands on than he needs to be, and not more hands off that you'd want him to. With Kim, there is always an open door, a perfect balance and an ideal kind of support that comes with understatement and self-deprecation, and with a sharp mind and a steady hand.*

Execution and Scaling

Success depends on the imagination of the idea, the importance of the problem, the elegance of the solution, the quality of the underlying magic or secret sauce, the competence of the sales team, the capabilities of the extended product, the pull of the marketing materials, and the adequacy of the financial resources. But it also depends upon a sometime intangible process called *execution*.

Execution is getting things done, making things happen, turning the vision into reality. The management consulting firm Gartner has defined the "5 pillars of strategy execution" as follows:

"Strategy formulation
Planning
Performance management
Strategy communication
Organizational capacity".

I would add "course correction" and list them in a very different order.

The first step in execution is strategy formulation. What are your goals? Over what time frame? What are the desirable outcomes? What are the risk factors? These are high-level questions; more detailed tactics are discussed in the next pillar.

Planning comes next, an enumeration and elaboration of the tactics to realize the strategic goals. What are the tasks? How can they be carried out? What steps must be completed as prerequisites for launching other steps? What staff and other resources will be required?

Organizational capacity need be considered concurrently with planning. If the tactics require two experts in machine learning, or one very experienced digital marketing person, you need to be aware of the time to recruit and hire such a person, or you must engage consultants. In terms of hiring more software developers, keep in mind Fred Brooks' Law: *"Adding more people to a late software project makes it later."* Although this may not apply if your time frame is 3 years, it is undeniably true if you need the software in 3–6 months.

Next comes strategy communication. It is essential that all staff who will be involved, and by "all" I also include administrative personnel, understand and buy in to what is required by when and what their roles and responsibilities are.

A difficult part of execution is performance management. Even if planning has been detailed and thorough, it is difficult to decide what to do if development is much slower than anticipated or a sales cycle is longer than predicted or competitors suddenly announce new products replicating your key features.

Reacting to such situations almost always involve course corrections such as adjustments to development or sales or marketing tactics, and sometimes even to an entire strategy. These changes provide severe tests to the vision, credibility, and mentoring capability of a leader, as well as to the flexibility and adaptability of a management team.

One of the most difficult aspects of strategy execution is the need to *scale* the venture. Unless your goal is a lifestyle business, you need to grow, so you must develop a success strategy which works as you move from tens of customers to tens of thousands to tens of millions. Scaling is growth without corresponding increases in expenses, enabling increases in both revenues and profitability. The best examples of scaling are firms whose internet presence has gone viral with exponential growth such as LinkedIn, Twitter, Airbnb, Chess.com, Canva, and Wordle.

Hiring, Diversity, and Supervision

Startup success depends upon not just the management team but also the software developers and marketing, sales, and support staff who do most of the work. It is critical that you hire only the most talented and committed individuals, as even a single person of insufficient competence or weak energy can do huge damage to productivity and morale. Despite the limits to growth imposed by being unable to find enough good people, you must not lower your standards because you are in a rush.

Often overlooked as a major heuristic in hiring is *diversity*. There ate many reasons for building a team of people of different backgrounds than the founder. The first reason is excellence. As I think of the hundreds of individuals who have worked for me in either my role as a professor or that of entrepreneur, well over half have been women or have come from a culture and background very different from mine. Diversity results in a multiplicity of points of view and understanding, very helpful in designing for customers of whom only a minority are white males. Individuals from groups that historically have been denied participation in a profession truly appreciate the opportunity to excel. Finally, opt for a diverse work force because it is ethical—it is the right thing to do.

Getting the right people starts with hiring them. There are three key techniques to doing this well. One is posing problems, in an interview or even in a written "exam", questions that force applicants to do in the interview some task representative of what they will do in the job. The second method is to have applicants interviewed by at least three and preferably four or five persons with whom they would have to work, ideally individuals who would be in a supervisory, peer, and subordinate position if they were hired. The third technique is to scrupulously reference check, which many startups fail to do due to haste or carelessness. A major error is crisis hiring, when there is a need for a person so urgently that standards are forgotten; often leading to disaster.

In hiring, be as open as possible about what you are seeking and what you expect. Consider personal qualities, and written, oral, and interpersonal communication skills, and not just programming and other technical abilities.

Once a person is hired, the management challenge is supervising, which is a mixture of goal setting, performance monitoring, open and honest communication, course correction, knowledge dissemination, and mentoring. Supervisors must have good interpersonal communication skills and be sensitive to the emotions, affect, and self-image of the people working for them.

Have regular performance reviews in which you give ample praise but also suggestions for improvement. Ensure that your "door is always open" to assist with problems; provide mentoring at every available opportunity. Develop standards for fair promotion and salary increases; make sure they are understood by staff and are not mysterious. If someone is not performing, give them ample warning, and try to help them succeed, giving them ample opportunity to improve. But document your concerns in detail, which will make necessary terminations as open and honest as possible. Firings (this was always the hardest task I had as a manager) and layoffs are always painful but may at some point be necessary.

Tech startups that recruit and interview thoughtfully will rarely have to fire people. Yet sometimes there will arise a need to reduce the workforce through layoffs because of a change in strategy, when a product line or channel to market is abandoned, or because the company is running out of cash. If you are contemplating some cuts, **cut deeply**, as you do not want to have to repeat the process in a few months. Once you do layoffs or miss a payroll, people dust off their resumes; if it happens again, they actively seek other jobs to have some options; if it happens a third time, the good ones will be gone within a month.

Firms differ in their policies when someone is let go. Some ask terminated employees to clean out their desk, then escort them to the front door, and cancel all electronic privileges that same hour. This may be reasonable for a firing based on poor performance, but I believe that cutbacks for corporate reasons should be done as gently as possible, allowing those remaining to grieve with those that have been let go.

Fairness and Generosity with Compensation

As we have seen with the cases of the CEOs of SideFX and Caseware, employees value vision and humanity in their leaders. Being generous, sensitive, empathic, and fair reaps rewards by encouraging maximal output from a team and keeping employee turnover low.

Yet vision will only take you part of the way. Compensation is a critical issue for startups. Those founded by experienced entrepreneurs with good track records and with strong financial positions (Caseware and Nuula are examples) can afford to pay top dollar to get the best people. Yet most startups cannot do this at the beginning.

As discussed in the previous chapter, founders may have to forego or defer salaries during the initial months or a year or two until significant capital is raised or revenues start flowing. They draw much or most of their compensation from the future value of their founders' stock. The same may hold true of the startup management team. but their equity positions will be smaller so they will have to be paid near-market salaries.

There will be times of great stress for engineering staff and software developers in a startup. People will work long hours, including most evenings and many weekends, to get the big demo ready for a tradeshow, or to ensure that a new release of a product is delivered to customers on time. You will likely not have the funds to pay for every hour of extra work, and certainly not for overtime pay. After the crunch is over, celebrate, and allow people ample time off to make up lost family time.

There will be difficult times, sometimes due to economic or competitive forces that are no fault of your firm. Although you may not want to share every little hiccup in your fortunes, being open and honest with the team about a grave peril is the right thing to do and in the long run the best strategy to take. You come across as a mensch. Your employees may surprise you with their commitment, energy, and great ideas. The case of Kim Davidson and SideFX in the early 2000s is a good example.

Being an executive or manager in a high growth startup is very challenging. Close relationships and openness within the management team are essential, including having difficult conversations about everything including duties, responsibilities, compensation, decision-making, and control. It also can be very lonely, so having some competence in other portfolios increases the depth of the dialogue and the quality of the resulting decisions.

All employees, including administrative and clerical staff, as well as all board members, should have a stake in a tech startup with a stock option or a profit-sharing package.

Profit sharing works well once a startup reaches maturity and is making a profit. 5–25% of a firm's earning are earmarked as profit to be distributed to employees based on a combination of their salary level and their performance that year. Caseware International did this for many years to good effect. But for many companies, at the beginning, and often for years, there are no profits, so the desirable technique is to grant *stock options*.

The basic idea is simple. Let's say your stock is worth 20 cents a share (valuation is beyond the scope of this book). A grant of 1000 options gives the employee the right to buy shares at that price in the future. All option holders have an incentive to increase the value of the company, let's say to $5 a share in 4 years, so that they can buy the stock for $200 and sell it for $5000 once the shares and be traded, or once the company is acquired and the shareholders are paid for their stock.

A stock option plan often has the options vesting over time, with a typical period of 4 or 5 years. This means that the right to purchase the stock only arises in stages, a fraction per year. Such as plan is often described as *golden handcuffs* for an employee, as it provides an incentive for them to continue to work for many years at the company.

There will be situations where it may be financial advantageous to outsource work to regions of the world where salaries are much lower. In deciding if you should do this, keep in mind the subtle but substantial coordination costs due to time zone and linguistic challenges, as well as your ethical responsibilities to your town, state, or country to support your economy with good jobs. Shaw Industries is an excellent example of a growing firm that has reaped rich rewards by continuing to build employment in their community.

You may be uncertain as to how to interact with competitors, whom you will likely meet at conferences and tradeshows. Know what information is secret and confidential; be open in sharing and discussing what is public about your firm and its products. The relationships you build in such circumstances may be invaluable when you are hiring or when you are looking for your next job. They may also serve as foundations for future strategic alliances that you cannot anticipate today.

Boards of Directors

Thus far, I have discussed desirable attributes and jobs of the leadership and management team, and of the staff. Wise startups also recruit, nurture, and listen to the advice of a Board of Directors or Advisers.

The Board of Directors has critical legal responsibilities. It is elected by the shareholders. Its function is to look out for the financial interests of the shareholders. Its primary tool in this regard is its obligation to appoint and sometimes remove the CEO. A wise CEO will reach out to board members for counsel on critical decisions, to get their impressions of the performance of other members of the management team, to make introductions with a view towards strategic partnerships, to use their network to help in recruiting new members of the management or senior staff, and to help them envision futures for the industry and the company. Board members should also not hesitate to speak privately to the CEO if they feel that s/he has gotten off the right track or is contemplating what they feel is unwise or unduly risky.

Members of the Board are also legally responsible for the costs of unpaid salaries in case the company fails, so many individuals are not keen to take this on for a startup, especially for those that are undercapitalized. Hence an alternative is to appoint a Board of Advisers, which is an informal body with no legal status or liabilities but with the role of providing wise counsel to the CEO and other senior executives.

Cast a wide net when thinking of board members, with the goal of getting a mix of industry and business knowledge as well as some entrepreneurial background and ideally links to sources of capital. Board members should be compensated with options vesting over time.

University Partnerships and Incubators

It is unlikely that you will be able to recruit a management team and hire a staff that has every skill, connection, and business leverage that you will need. Strategic partnerships help you fill these gaps without the need to increase your head count faster than wisdom dictates.

If your vision and your technology is leading edge, as it should be, partnering with universities and university professors is a great idea. Professors and students have ideas which can be commercialized. Their labs can do research, such as evaluating the user experience of your product. Many government programs provide financial aid for industry-university partnerships. Students form a pipeline of new talent for a company, as has been the case with the University of Toronto for SideFX, Caseware, Winterlight and Braze, and has long been the case with Canada's University of Waterloo and Microsoft, with its coop program often being the greatest source of new hires for the firm.

Universities as well as industry trade organizations now often have resource centers known as *incubators*. These provide seed capital or facilitate access to seed capital and offer low- or no-cost office space and mentoring to new venture founders and start-ups. The most successful of these are YCombinator in Silicon Valley and the Creative Destruction Lab of the Rotman School of Management at the University of Toronto, but there are now many of them in many geographies. For example, the University of Toronto has close to a dozen operating in various faculties; there are also well over a dozen more in the greater Toronto area.

Business to business partnerships occur in variety of forms. *Sales representatives* find customers for a firm's product and receive a commission in return. A sales partnership with closer technology and market cooperation is the relationship between a firm and its *distributor*. The distributor provides knowledgeable access to a particular market. For example, Caseware achieved global expansion without the need for major infusions of capital by finding small entrepreneurial firms in countries such as Germany, England, Norway, and South Africa that had accounting, accounting, and software skills, as well as knowledge of the local culture, to represent them in that country. Selection was done with great care. The relationships were deepened by senior executives during their six-week round-the-world trips twice each year.

Firms also make technology sharing and cross-licensing partnerships, for example Apple and Microsoft so that the latter's software was available on Macintosh computers, and Adobe and Linotype so that the Adobe could use Linotype typefaces in its products.

Ethical Responsibilities

In summary, we have seen how critical are the leadership and management skills of a founding team of entrepreneurs. It is also essential that they keep in mind that their stakeholders go beyond the management team, employees, board, and shareholders to include customers and citizens of their country and the world. Entrepreneurs have ethical responsibilities and need a moral code and sense of right and wrong. This is the subject of the next chapter, where I highlight many of the ethical challenges already discussed via the decisions and actions of several startups.

V. Blue Rock Therapeutics

Born and raised on a Saskatchewan farm, Gordon Keller completed a PhD in immunology, then worked in Switzerland, Austria, and the United States prior to returning to Canada in 2007 to establish the McEwen Centre for Regenerative Medicine in the Toronto Medical Discovery Tower. His research focused on using embryonic stem cell technology to isolate developmentally significant stem cells which can then differentiate and be programmed into a variety of cell types. These cells can then be used in regenerative medicine approaches to heart and blood-cell diseases.

Blue Rock Therapeutics was founded in 2016 to provide an avenue for commercialization of the results of the research of Keller and other scientists by the Versant venture capital firm and the pharmaceutical giant Bayer AG. One of their greatest successes was an ability to grow heart cells in the lab, and to begin work seeking to use them to increase the effectiveness of cardiac transplants.

Blue Rock has had a parallel line of development, aimed at developing pluripotent stem-cell derived dopaminergic neurons for patients with Parkinson's disease, with results that are now in Phase 1 clinical trials to evaluate safety, tolerability, and preliminary efficacy.

Bayer AG fully acquired Blue Rock in 2019, ensuring that it would have the resources to do the work over the very long research, development, and clinical trial validation phases required to develop advanced medical technology products.

W. Braze Mobility

Pooja Viswanathan spent 2006 through 2012 working on a doctorate in computer science, specializing in AI and computer vision. Her research was designing and developing an intelligent powered wheelchair that aids wayfinding and collision avoidance for seniors with cognitive and visual impairments. After continuing development and evaluation of such technologies and business models for making them commercially available as a post-doctoral fellow at the Toronto Rehabilitation Institute, she founded Braze Mobility in 2016.

Braze Mobility's original vision was the development of "smart wheelchairs" that could guide themselves relatively autonomously and do navigation and collision avoidance. Yet Pooja learned in work with and interviews with real wheelchair users and observations in environments in which wheelchairs are used (such as hospitals and senior care homes) that seniors wanted to be in control. She also saw a way to enter the market with less need for investment and without the need for regulatory approval of her product as a "medical device". (This pivot from trying to design and build smart wheelchairs to designing and building sensors which can make wheelchairs "smart" was discussed in the previous chapter.)

Hence Braze's initial products are blind spot sensors which digital perception and control logic that transform wheelchairs into "smart wheelchairs". Braze seems to be moving forward well with technology development, user experience design, and customer success, and using influencers and partnerships to achieve early market penetration.

X. Nuula

Mark Ruddock has almost 40 years of experience leading successful high-technology ventures, operating internationally, and achieving successful exits in the financial services sector. After being recruited to be CEO of BFS Capital in 2018, a financial products and services firm that had issued over $2 billion in loans to over 23,000 companies in over 400 industries, Mark orchestrated BFS morphing into Nuula in April 2021 with a new financing of $20 million in equity funding and a $100 million credit facility.

There are over 7 million small businesses with 2–50 employees and over 30 million owner-operated sole proprietorships in the United States. Nuula is developing a suite of apps to enable small business owners to monitor critical financial, payments, and e-commerce data, as well as to access financial products including a Line of Credit enabling small loans to bridge cash flow valleys. The apps are integrated into one super-app with a common user interface, providing information about the cash, credit, and reputational status of the business. The interest rate for the loans is easy to grok—1% per day for each $1000 borrowed, which makes sense if the need is short-term, especially since banks in most countries are unresponsive to the needs of small business operators.

Ruddock recently orchestrated a major pivot for Nuula, resulting in the strategy change that it will in the future outsource the provision of credit and loans. The benefits of this strategy were discussed in the previous chapter.

Ethics

Good business decisions

Responsibilities to stakeholders

Stakeholders

Shareholders   Customers   Society

Employees   Competitors

R. Baecker, *Ethical Tech Startup Guide*, Synthesis Lectures on
Professionalism and Career Advancement for Scientists and Engineers,
https://doi.org/10.1007/978-3-031-18780-3_8

Technology is neither good nor bad; nor is it neutral.
Melvin Kranzberg (1986). *Technology and history:*
Kranzberg's Laws. Technology and Culture 27(3):
544–560.

You don't have to choose between being scientific and
being compassionate.
Robert M. Sapolsky (2017). *Behave: The Biology of*
Humans at Our Best and Worst. Penguin Press.

Three of my students created a startup to commercialize novel mobile software to aid
people with speaking challenges such as autism spectrum disorder. The startup did not
create a board, a critical error. Two of the founders left the firm, one to earn much more
at a digital media giant. Despite advice to raise capital, they refused to do, wanting to
keep their share percentages as high as possible. The CEO worked very hard for three
years and found 2000 customers for his software. One day, by accident, I discovered that
he had abandoned the firm and moved to California to participate in a startup with large
venture funding. Nobody was told, not the one employee who did sales and support, not
me as the originator of the concept, likely not even the family and friends' investors. The
CEO promised to help me find a firm to purchase the startup but did nothing to make
this happen. It is not unethical to get burned out and need to seek a new position. Yet, in
abandoning without adequate notice the company's shareholders and other stakeholders,
mostly sadly the customers, the CEO behaved unethically. For years afterwards, I would
get sad emails from parents of children who had used the software to help them speak but
now needed bugs fixed.

There are other better-known cases of unethical behavior by founders, in recent years
most notably with Uber and Theranos. This chapter discusses issues that have arisen in
our case study firms as they have sought to create and sustain an ethical tech startup.
Management must also continually keep in mind that they have ethical responsibilities
not only to corporate shareholders but also to employees, customers, and competitors,
and to society as a whole.

Ethical Corporate Behavior

I shall begin by summarizing ethical corporate behavior and challenges previously discussed.

Successful startups exploit technological opportunities relevant to societal problems
and create solutions to these problems, bringing value of customers. I have presented
examples of firms with "ethical missions" seeking to improve health, learning, community, and the environment.

Successful startups have solutions embodying an underlying magic which helps
differentiate them from competitors and is a source of competitive advantage. Ethical

solutions must work reliably and consistently and be safe. The Therac-25 and 737 Max cases illustrate the damage and loss of life that can result from unsafe technologies.

Successful startups prioritize the customer user experience, seeking to ensure that the products are as easy to learn and use as possible, that they are comprehensible, and that their use is normally error-free.

Successful startups know who they are and what their values and strengths are, which enables them to project and identity and make promises to customers that they can keep.

Successful startups have viable business models and place emphasis on treating their customers ethically with respect to communications, pricing, and customer success.

Successful startups manage their finances intelligently making use of forecasting models; they are scrupulously honest with respect to financial reporting and communications with stakeholders such as shareholders.

Successful startups are led by leaders and management teams who have vision, intelligence, drive, and optimism; who create a vibrant and collaborative culture; and who treat all employees with openness, fairness, and generosity.

Ethical Opportunities and Challenges

I have written extensively in recent years of numerous examples of firms that began as visionary and entrepreneurial but whose products have had evil consequences, mostly unforeseen by the founders. In the domain of digital technologies, these consequences include technology addiction, cyberbullying, hate speech, disinformation, loss of privacy, inappropriate surveillance, cyberterrorism, unsafe technology use, rapid automation without adequate social safety nets, monopoly control over industries, and premature, risky use of unreliable AI.

We **can** do better. The job is about more than building something nifty and becoming rich. Opportunities for "ethical tech" startups are everywhere, as the world is beset with problems. The globe is becoming environmentally unsustainable. What can you do to slow global warming? Hate speech on social media is spreading evil and costing lives. Can you find better ways to detect it automatically? Self-driving cars are inevitable given humanity's proven inability to drive safely. Can you help make autonomous vehicles reliable and safe? There are major incentives for governments and corporations to adopt AI algorithms before they have been proved to be reliable. Can you help make AI decisions and actions explainable, fair, and trustworthy, with the appropriate people and firms held accountable and responsible?

All our startup firms were created with noble goals such as those listed above. Yet, despite the loftiest of aims, startups often encounter challenges in behaving ethically, usually because of the desire for market dominance and greater profitability, sometimes because the original vision is subverted by actors pursuing aims that are self-serving or malicious, and sometimes because their operations impact different stakeholders in different ways. Microsoft, Apple, Twitter, and Airbnb are excellent examples.

Case Studies of Ethical Challenges

Microsoft was known as the Evil Empire in the 1990s due to its success, its bundling of Explorer in with Windows, thereby dooming Netscape, and in part due to the abrasive style of its leaders Bill Gates and especially Steve Ballmer. In 2000, a judgment was handed down calling the firm "an abusive monopoly, although the case was settled out of court with the U.S. Department of Justice in 2004.

The "evil" label was not totally fair, as Microsoft's technology had brought much good into the world, increasing productivity, and spreading the availability and benefits of computing widely. The firm has always been enlightened in its support of university research, its creation of the world's top digital technology research lab which it distributed across many locales including India and China, its openness in publishing research results, and its encouragement of interaction between the company and academia.

Yet recently, under the leadership of CEO Satya Nadella and President Brad Smith, it has pivoted to a kinder, gentler style, and moved aggressively to be a force for good. Microsoft pledged and has mostly invested $750 M to support affordable housing in the Seattle area and invested $115 M in nonprofits working on climate change and technology accessibility. It has worked towards global regulations dealing with cybersecurity and facial recognition; and recently phased out the use of its software intended to recognize human emotions but had been criticized as being flawed and unscientific. Microsoft has also released the second version of its Responsible AI Standard, and proposed a Hippocratic Oath for coders, aimed especially at those working in artificial intelligence, to be like that for doctors—"first, do no harm".

Apple Computer's superior learnability and usability has done enormous service to the world by empowering over a billion people with technology they can master. Its early commitment to user experience research helped launch the discipline, although its prominence in the field diminished quickly after the return of Steve Jobs, who disbanded the research lab. The company has also long supported data privacy through a variety of public stances and as well as the design of its software. It also seems committed to environmental sustainability.

Apple's success has been due in part to its walled garden, its insistence on controlling what apps run on its hardware. This has enabled greater consistency of the user experience, an aid to learnability and usability. The commissions it charges for use of the iPhone App Store are widely viewed as unreasonable. Its argument that you do not have to distribute on iPhones if you don't like our fees is self-serving, since it and Google have a monopoly, with Apple taking the lion's share in 2021 with $85 billion in revenue as compared to Google's $48 billion. Apple has also used its control of the environment to hinder apps competing with Apple apps to run on its iPhones, and has frequently developed its own apps after it certain third-party apps become successful on the iPhone.

Apple has been justly criticized for the tax practice of using its operations in Ireland to avoid tens of billions of dollars in taxes, although legal proceedings are underway that would recoup some of them. There is also evidence that it has been insufficiently diligent in forbidding exploitative and unsafe labor practices in the manufacturing of its computers, especially in China. More generally, its strong commitment to a partnership with China, and its reliance upon China both for manufacturing and as a source of revenues, especially for iPhones, has led to a morally questionable alliance.

Twitter has long had to make ethical choices because of the tension between the commitment to free speech and the increasing prevalence of hate speech and disinformation on social media platforms. Does the harm caused by evil speech—violence against minorities or poor health outcomes caused by medical misinformation—warrant removal of the messages in what seems to some as censorship? For a long time, Twitter tried to take a balanced approach, labeling some tweets as "potentially misleading" or "glorifies violence". The problem was made more difficult because one of the worst offenders was Donald Trump, then President of the United States. After the riot at the US Capitol on January 6th, 2021, which was encouraged and later praised by Trump's tweets, his account was at first suspended and then permanently suspended. This action was both widely praised in the interests of safety and condemned as a violation of free speech.

Twitter has also been in the news in 2022 because tech entrepreneur and tycoon Elon Musk bought the company. Musk has 81 million followers, making him one of the 10 most visible people using Twitter. He has been known to tweet strong opinions and statements critical of many people. His tweets and public statements in 2022 insulted and disparaged Twitter's product, management, and content policies. Musk has made clear he will make many changes to Twitter, ranging from the innocuous improvement of allowing the editing of tweets to having Twitter rebuilt using the Web3 paradigm to making free speech the dominant value to the detriment of reasonable content moderation.

It is therefore reasonable to fear that the future of the service will no longer be as a vehicle for reasoned constructive dialogue but instead as a forum for even more disinformation and hate. The egotistical, unethical actions by one of our age's most brilliant technical entrepreneurs have driven down the stock price, decimated the workforce and especially those working on content moderation, and damaged, possibly fatally, one of the internet's most interesting and socially relevant companies.

Airbnb's founding idea was to provide travelers a new form of accommodation that would be more congenial and inexpensive, provide homeowners extra income from unused space, and enriching the lives of both through making new friends. Because this is a new form of accommodation, quality and safety standards have been developing slowly, with well-publicized incidents when guests or hosts do not experience what they expected and believe they were promised. Given that Airbnb is yet to achieve consistent profitability and also sustained significant setbacks during the first year of the pandemic, it is unclear whether the company is investing sufficiently in screening hosts and guests, investigating complaints, and acting on them. Like so many firms in a capitalist economy, it misleads prospective guests by hiding the full list of charges (including cleaning fees and Airbnb's commission) until the last moment when they are about to make a booking.

Yet the deepest challenge to ethical behavior arises from property owners and landlords perverting Airbnb's founding vision. Sensing quite correctly an opportunity to make a great deal of money, individuals have taken long-term rental properties and repurposed them for more profitable short-term rentals. They also have bought up vacant apartments and turned them into Airbnb rental units. This has led to several bad consequences. Their goal is primarily profit, so they invest little in quality furnishings and human support of and interaction with their guests. They do little or nothing when the renters (the word "guests" no longer applies) cause disruption to a building or neighborhood with loud parties. Their demand for properties also makes housing and long-term rental accommodation in a neighborhood much less affordable. Many municipalities, such as Barcelona, Rome, and Lisbon, have enacted regulations in response to these bad side effects of what was once a noble concept.

Is Airbnb acting ethically? I hope that quality control is improved and that bookings can be restricted to homeowners with extra space—the founding vision—and that "Airbnb cut-rate hotels" will be forbidden. Yet, given the stock market's unrelentless pressure for growth, I am not hopeful that the firm will pivot back to its founding vision.

In summary, despite the many problems caused to society by certain technologies or how they are being used, there are enormous opportunities to incubate ideas, develop them into innovations, and commercialize them via ethical entrepreneurial ventures into products and services that can lead to a better world.

Building a successful tech startup requires constant effort and imagination along with bottomless reserves of energy. Building an ethical tech startup requires thinking ahead and constant vigilance along with bottomless reserves of resilience. This book has featured cautionary tales and inspirational stories of founders and companies who have struggled with this balancing act.

Just like privacy is best ensured by design, so are ethics. Strive to build your product and your company ethically-by-design from the very beginning. Enroll your whole team in a mission to invest the future in a sound and ethical manner. You will then be more likely to build a company you can be proud of not just because of the future it brought to life, and not just because of its financial returns, but also because of how it acted and the choices it made along the way.

Y. Drisit

Late in 2020, in the depth of the pandemic, which had been disastrous for the travel industry, Alex Backer, the founder of QLess, started Drisit with two partners, one with 30 years of commercial strategy and business development experience, and the other a senior software development engineer with 25 years of experience in areas including machine learning.

Drisit, a clever contraction implying visits by drone, will allow users to rent a drone for short periods of time to see and explore and learn, ultimately in many parts of the world. Drisit requires partners to supply destinations or customers, such as tourist attractions wishing to expand their flow of visitors, educational institutions or educators seeking to expand the horizons and experiences of students, or utilities seeking to do remote inspections of the devices or systems that install over wide geographic areas. Owners of destinations specify the range of flights paths as well as geofencing to enforce limits where necessary.

Because Drisit opens new avenues for imaginative markets and business models, other opportunities come to mind. Would people use it to attend birthday parties of grandchildren in far-flung locales? Alex Backer believes that the Drisit drone economy could be as significant as the telephone. The phone enabled people to speak and hear anywhere. Drones can let us see anywhere. What applications can <u>you</u> think of that could lead to a better world?

Z. Wordle

In 2021, Josh Wardle, a Brooklyn, New York, software engineer, created a simple online word game for his partner as a token of his love. Players of Wordle are given 6 tries to guess a 5-letter word; on each try, they are told which guessed letters are in the correct position, which appear but in different positions, and which are not in the target word.

Over several months, the couple and other members of his close family played, followed soon by his WhatsApp extended family. The game was so popular that he published it online in October. On Nov. 1, there were 90 players. By Jan. 2, 2022, there were over 300,000 players, in great part because of its viral component that allowing users to notify friends about the game by sharing a graphic of their success with that day's puzzle. The Wordle phenomenon was written up in the New York Times on January 3. Less than 30 days later, the paper bought Wordle to add to its Games section for an amount said to be in the "low 7 figures".

This case is unusual in that the time from idea conception to software creation to marketing success to founder exit through selling the product was under a year, although it is indicative of modern internet speed. The sale was a win–win. Wardle was no longer having fun building Wordle as a business. He was able to realize far more than he ever imagined and continue his chosen life as a software developer, while preserving Wordle's free status and keeping it free of advertising. The Times reported in its May 4, 2022, Quarterly Report: "Wordle brought an unprecedented tens of millions of new users to the Times, many of whom stayed to play other games which drove our best quarter ever for net subscriber additions to Games."

Wordle has even spawned many competitors, and those using the same name have been asked to shut down by the Times.

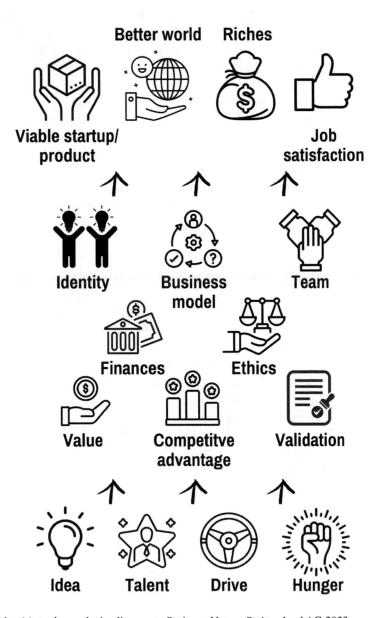

R. Baecker, *Ethical Tech Startup Guide*, Synthesis Lectures on
Professionalism and Career Advancement for Scientists and Engineers,
https://doi.org/10.1007/978-3-031-18780-3_9

Here is a summary of my lessons to guide new entrepreneurs, to enable them to harness their ideas, talent, drive, and hunger for success to create dynamite products and viable startups, to achieve job satisfaction and even riches, and to help make a better world.

Seek solutions to important problems. Develop competitive advantage through proprietary technology and/or distinctive competence. Validate your goals, strategy, and intuition by user experience design at every stage from product conception to product shipment. Think about who you are, and what your strengths are; leverage it in your strategy and communicate this to customers. Design your business model and go-to-market strategy carefully; devote significant resources to digital marketing. Plan and project your finances carefully; raise funds expeditiously but vigorously. Hire only the best people; motivate and guide them with inspirational leadership by a broad and capable management team. Look for ethical startup opportunities; manage your company and team ethically.

Let me elaborate on these points by summarizing principles for success.

1. There are many reasons to create a company, and also many not to start one. Realistically assess before you begin whether it is the right thing for you to do so **now** with your current idea.

Value

2. Build your startup on *innovation*—creating novel ideas, processes, methods, or products.
3. Innovation must be harnessed by *entrepreneurship*—creating, nurturing, and growing a successful and sustainable new venture.
4. Many innovations are responses to technical or social *opportunities* to do things better, faster, or cheaper, or to do something totally new.
5. The most significant innovations are responses to important societal *problems* causing *pain* to many individuals or corporations.
6. Innovators must develop viable ways to exploit opportunities and to develop *solutions* to substantial problems.
7. Startups are well advised to address and *focus* on small but substantial *niche markets*.
8. Be confident that *saying "no"* to other opportunities will allow you to develop a strong market position in your niche market from which you can later expand geographically and/or with follow-on products.
9. Anticipate and plan what these follow-on markets or products could credibly be.
10. Try to anticipate what negative consequences might arise from your product and see if you can develop a strategy so that they will not occur.

11. You must *size* the niche market and ensure that it (given a realistic product price) will allow you to build a business of adequate size.
12. Timing is critical. Many products have *market windows*. It is unwise to be too early, but you need to be one of the first to enter the market.
13. Entrepreneurs must articulate *value propositions* for their products, making clear qualitatively and ideally quantitatively why their solutions work well for customers and solve their problems.

Competitive Advantage

14. Big technological opportunities and societal problems attract competition. To succeed, you must develop significant *competitive advantage* with respect to the competition.
15. The strongest competitive edge comes from *proprietary technology*, products or processes that you have that your competitors do not have.
16. You must guard your proprietary technology with *intellectual property protection* methods such as trade secrets, copyright, patents, and trademarks.
17. Just as important is that you have a *distinctive competence*, expertise which you have that your competitors do not have, which you can exploit for competitive advantage.
18. Be first or one of the first to address a market segment to obtain a *first mover advantage* with respect to others who come later.
19. A first mover advantage and more generally a strong market position creates *barriers to entry* which make it costlier and more difficult for others to compete.

Validation

20. Many good ideas and innovative solutions do not yield successful entrepreneurial ventures, so methods of *validation* are needed to calibrate the likelihood of success and increase its profitability.
21. *Competitive analysis* helps one understand your position with respect to competitors and suggests how to improve that position.
22. *Differentiation* of your company and products from competitors suggests what strengths to exploit and emphasize in your *positioning*, and helps customers understand why they should buy your products.
23. Almost all products affect multiple *stakeholders* beside the primary users; consider in advance how your product will affect all stakeholders.
24. Creating *personas* of these stakeholders helps you in the product design process.
25. A further refinement of a persona is a *use* case describing how the product will used. Enabling efficient, comfortable, and safe use is a critical part of the design process.

26. *User experience* research, including both ethnographic field research and laboratory usability research, helps you understand your users, if and how your design is good, and how to improve it.

27. *Quality assurance* to ensure that critical flaws have been identified and removed is essential prior to product shipment. One QA method is *dogfooding*, the active use of your software by members of your team.

28. A focus on *customer success* once your product is being used by customers is essential for building customer loyalty and for learning how to improve your product.

Identity

29. Startups must understand their *identity*—who they are, what they are good at, and what they have that is special that could be the basis of a successful venture.

30. Identity must be represented and displayed to potential customers through corporate identification via a memorable name, a clever logo, a succinct mantra, and communicative vision and mission statements, all contributing to a *brand*.

31. Essential for raising funds, attracting customers, and recruiting employee is a succinct *elevator pitch* which communicates your identity, the essence of your product, and your competitive advantage

32. Also essential for raising funds is a concise *pitch deck* anchoring presentations you make to venture capitalists; this also serves as a summary of your strategy which management can review and improve.

Business Models

33. Startups must be able to articulate a *business model*—how it will use people and resources to make money. It is helpful to diagram this on the *business model canvas*.

34. A critical aspect of the business model is the *go-to-market plan*, which describes how sales and marketing activities will be used to launch a product.

35. Design your *augmented product* to surpass customer expectations; in the case of digital technology, this will likely include both algorithms and data. But first develop and ship your *minimal viable product*.

36. Both Business to Business (B2B) and Business to Consumer (B2C) ventures must design pricing and *revenue models* carefully with respect to your competition, your product's value to customers, and your costs.

37. A *freemium* revenue model, in which customers can use a subset of your functionality at no cost, has proved to be valuable for many vendors of internet apps.

38. Modern tech ventures must master *internet marketing* including the use of social media and search engine optimization.

39. Know who the *influencers* are who carry weight in your marketplace. Work hard to be known and respected by them.

Finances

40. To understand the financial health of your venture, and to raise additional finances, *financial statements* of profit-and-loss, cash flow, and balance sheet need to be prepared regularly.
41. Most important in understanding your financial health are *key indicators* such as cash balance, cash burn rate, rate of customer acquisition, customer churn rate, revenue growth rate, and profitability.
42. You must create a *financial forecasting model* to predict key indicators. You need this to show to potential investors, but even more importantly, it is an essential management tool that forces you to make explicit assumptions about your venture and helps you to improve them.
43. Startups need to raise finances for product development and marketing. Entrepreneurs need to be aware of sources including government grants and tax credits; and sweat equity, family and friends, angel, and venture capital sources.
44. Unless you are planning to have a modest lifestyle business, startup founders need to anticipate stiff competition for market share and need to plan *growth strategies*.
45. Investors in your startup are your partners. You must plan for *exit strategies* so they can realize returns. Your founding partners should also anticipate and think about their own personal exit strategies.
46. Be scrupulously honest with yourselves, your employees, and your shareholders about financial crises and key challenges when they arise.

Team

47. Startups need *leadership*. The best leaders have superior expertise, vision, passion, and humanity, and have the inner resources to summon and convey confidence in the toughest of times.
48. Strong and visionary leaders model ethical behavior and apply their values to help establish a rich and positive corporate *culture*.
49. Leaders and management teams must get things done, make things happen, and turn the vision into reality. This is known as the art of *execution*.
50. The most successful companies can *scale*, that its, to grow customers, revenues, and profitability without corresponding increases in expenses.
51. A critical decision for leaders and a management team is *pivoting* a startup's strategic direction when the current approach is not yielding the desired product acceptance, market share, and financial returns.
52. Startup growth is only possible if it us run by a collaborative *management team* with complementary strengths in at least technology, product design and user experience, sales, and marketing.

53. Another key to success is a *team* of employees with extraordinary talent and commitment, ideally with increased motivation through owning stock options. It is important not to let critical needs for new employees and haste result in compromises on employee quality.
54. *Diversity* in hiring will increase your excellence; the multiplicity of perspectives will enhance your decision-making and the suitability of your products for different kinds of customers.
55. A startup's own resources can be bolstered through *strategic partnerships* with other firms that strengthen it in areas such as technology, sales, and marketing.

Ethics

56. The many critical societal problems such as environmental damage, devastating diseases and pandemics, and internet hate speech and disinformation lead to countless opportunities for "ethical tech" startups.
57. All tech startups should be **ethical** "tech startups", keeping values in mind in all of their dealings with employees, investors, and the public.
58. Always keep in mind your responsibilities to all stakeholders.
59. Treating members of your team openly, honestly, fairly, and generously is one major way for a startup to be ethical.
60. For the most part, companies are neither totally ethical or unethical; rather, they face on an ongoing basis a variety of opportunities for and challenges to behaving ethically.

It can be an exciting journey. You can create amazing products. You can change the world for the better. You can reap rich financial rewards. You can do this while acting ethically, although this will require to be certain of your values and rely upon these often as your business faces the inevitable challenges posed by growth.

Please realize that startups are not for everyone. To dive into a turbulent sea with raging waters and no life preserver and only a rudimentary ability to swim takes courage, commitment, a good idea, and the distinctive competence to give you a chance to succeed. There will be times when you must be willing to risk almost everything. Even if you are willing to do so at some point of your life, remember that it may not be wise at every point in your life.

I stress what I believe is the most dangerous trap for novice tech entrepreneurs, and that is to fall so deeply in love with your own technology that you are oblivious to what it will cost to take it from idea to shipping product, blinded and unaware of its flaws, and insensitive and unresponsive to the reactions of customers and potential customers. Tech startups are not about technology but about responses to opportunities and solutions to problems, all in service of users and customers.

No matter how good your idea is and how skilled you are, luck and serendipity will play an important role. We saw this with the origins of the mushroom material binding idea for Ecovative Design's product, the timing of the Democratic National Convention as an opportunity for Airbnb's founders, the unexpected article in TechCrunch which helped launch Chess.com, and the sudden emergence of the New York Times as a purchaser for Wordle.

I have chosen my 26 case studies with care, yet there are no guarantees for the future of these firms. For example, considering those five created in the past seven years (ignoring Wordle because it has already achieved a more than satisfactory exit for its founder and for the New York Times which purchased it), I am optimistic but in no ways certain that they will be able to meet the challenges they will face and to achieve the following results: Winterlight Labs will continue to develop innovative, even more powerful algorithms for dementia monitoring. Blue Rock Therapeutics will find its stem cell regenerative medicine solutions effective and safe through its clinical trials. Braze Mobility will expand its range of smart wheelchair solutions and achieve traction despite multinational competition. Nuula will find a way to get owners of small businesses, who historically have made little use of computers, to adopt their apps with demonstrable improvement in their management ability and financial results. Alex Backer's vision that people will want to explore the world virtually through their eyes and the control of drones will prove to be a brilliant idea for the future of Drisit.

I wish **you** well with your current or next startup!

Elevator Pitch, Competitive Analysis, Pitch Deck for This Book

The next few pages present an elevator pitch, a competitive analysis, a pitch deck for this book which follows the guidelines discussed in Chap. 4, and a script supporting the pitch deck.

It presents the competitive advantage of this book, a marketing strategy, the author's distinctive competence, and a financing plan and projected financial results, and concludes with a Call to Action for a publisher to accept the book for publication. Books do not get marketed to publishers with pitch decks, but this deck is a good and accessible example of the key concepts, which apply to all entrepreneurial ventures, including books, restaurant chains, and tech startups.

Note that the script follows a narrative so that it does not sound like isolated points. Although the slides and the script correspond, somewhat different material is presented in each medium. Most important is that the pitch ends with a call to action.

Here is the elevator pitch:

The Ethical Tech Startup Guide is the first book to present comprehensively yet concisely a set of principles for tech startup success, illustrated with modern examples and emphasizing the need for and challenges to ethical behavior.

Here is a competitive analysis:

Feature	Business model generation: a handbook for visionaries, game changers, and challengers	So What? Who Cares? Why You? The Inventor's Commercialization Toolkit	Ethical tech startup guide
Year published	2010	2006, 2015	2023
Innovation, problems, solution, value	+++	+++	+++
Underlying magic and competitive advantage		+	+++

R. Baecker, *Ethical Tech Startup Guide,* Synthesis Lectures on Professionalism and Career Advancement for Scientists and Engineers, https://doi.org/10.1007/978-3-031-18780-3

Feature	Business model generation: a handbook for visionaries, game changers, and challengers	So What? Who Cares? Why You? The Inventor's Commercialization Toolkit	Ethical tech startup guide
Validation and user experience	++	+	+++
Identity: defining it, communicating it	+++	++	++
Business models and go-to-market plans	+++	++	++
Finances and financing			++
Leadership, management, and team		++	+++
Ethical entrepreneurship			++
Financial forecasting model			++
Conciseness, ease of use	+	+++	+++
Author's international reputation	+++	+	++
OVERALL	++	++	+++

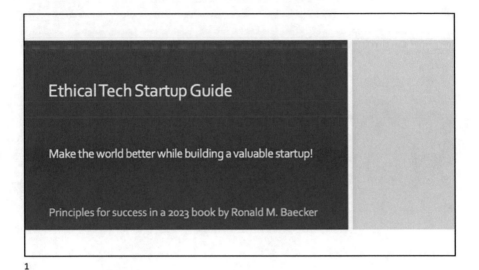

1

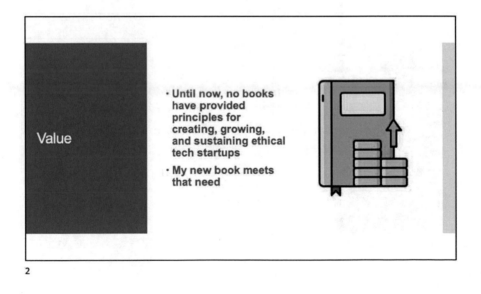

2

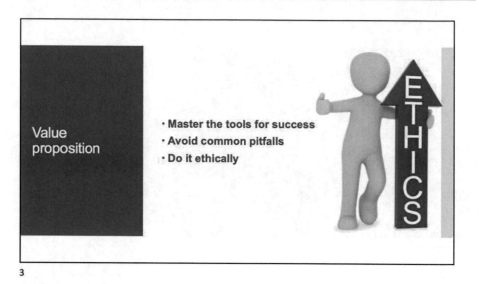

Value proposition

- Master the tools for success
- Avoid common pitfalls
- Do it ethically

3

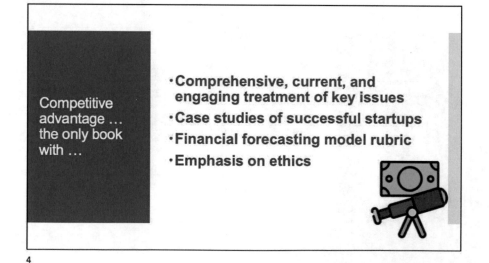

Competitive advantage … the only book with …

- Comprehensive, current, and engaging treatment of key issues
- Case studies of successful startups
- Financial forecasting model rubric
- Emphasis on ethics

4

Validation: competitive analysis

	Business Model Generation	So What? Who Cares? Why You?	Ethical Tech Startup Guide
Problem & opportunity	+++	+++	+++
Underlying magic, IP		+	++
Validation	+++	+	+++
Identity	+++	++	++
Business model, go-to-market	+++	++	++
Finances and financing			++
Leadership and management		++	+++
Ethics			++
Market position	Very strong	Moderate	Startup

5

Validation: an expert opinion by Andrew Silverman, MBA, CPA, startup mentor (NYU, Duke, MIT)

- *"For entrepreneurs building the next unicorn, or anybody interested in ... building great technology companies — The Ethical Tech Startup Guide is an amazingly rich and valuable resource ... comprehensive ... spanning the entire life cycle — from ... initial idea ... through scaling and growth.*

     What makes the book so valuable [is] viewpoints and insights from both business and technology perspectives. ... The book contains a trove of case studies of great technology companies and the keys to their success. ...

     ... emphasizes building ethical startups. As technology has recently been used to spread misinformation and conspiracy theories with startling speed and scope, causing mistrust and other damaging effects on society, this book is a welcome breeze and inspires hope for the future."

6

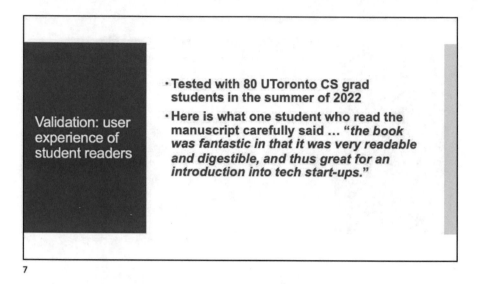

7

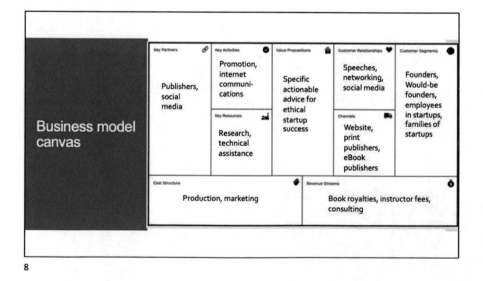

8

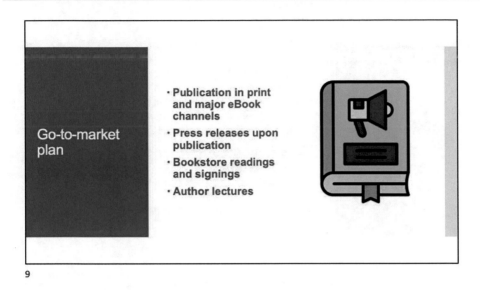

9

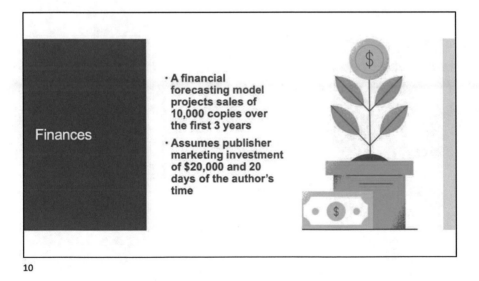

10

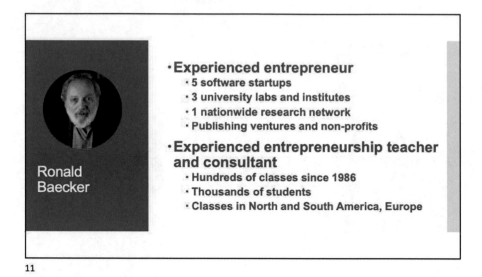

11

12

Script to Accompany the Pitch Deck for This Book

Good morning, publishers!

[Slide 1] In this troubled times, young people have two visions that animate their dreams of a good life. One is to create a startup and make a fortune. The other is to make a difference, to help make a better world.

[Slide 2] My new book, *Ethical Tech Startup Guide*, will be the only book that responds to both visions. There are no other books that provide practical and comprehensive principles for ethical startup success.

[Slide 3] Readers of this book who master the principles will have the tools for startup success. They will be aware of common ethical pitfalls.

[Slide 4] This book is unique is its grounding in 26 case studies of modern ethical tech startups, its provision of a useful financial forecasting model rubric. and its emphasis on ethics.

[Slide 5] This competitive analysis chart compares my book to two other leading texts for startups.

[Slide 6] I have distributed drafts of the text widely. Let me tell you what people have said: One seasoned industry veteran with 3 decades of experience and 2 successful startups under his belt, wrote: "I wish I had this to refer to early on. I faced all these issues over the years". The slide shows a more detailed commentary by an experienced businessman and startup mentor.

[Slide 7] Students using a draft of the text in the summer of 2022 were also full of praise. The slide shows one summary by a student.

[Slide 8] Publishing and marketing of the book is described by the Business Model Canvas shown on this slide.

[Slide 9] Key to success is an investment in promotion, including readings and signings at bookstores. I will be your partner in this effort, as I travel the world giving lectures, short courses, and tech startup bootcamps.

[Slide 10] I forecast that the result, given a modest investment in promotion, will be 10,000 sales over the first 3 years, which will then be followed by a second edition and accelerating interest and sales.

R. Baecker, *Ethical Tech Startup Guide,* Synthesis Lectures on Professionalism and Career Advancement for Scientists and Engineers, https://doi.org/10.1007/978-3-031-18780-3

[Slide 11] In closing, let me reiterate my entrepreneurship credentials, making me uniquely qualified to write this book. For five decades I have been an academic, an author, creator of five ethical software startups, and teacher of thousands of entrepreneurship students.

[Slide 12] This book will be judged as one of the best business books of 2023. Please see me so we can begin a discussion or write me at ronbaecker@gmail.com.

Thank you.

The Financial Forecasting Model Rubric

At the end of this Appendix, I show in 2 pieces (left, right, slightly overlapping) the forecasting model rubric discussed in Chap. 6.

Here is a guide to the model.

First, consider the rows of the spreadsheet. They are organized into sections. Rows 2–4 detail assumptions and about customers. Rows 6–12 detail assumptions about personnel in various job categories, which for many tech startups generate most of the expenditures. Rows 14–36 show the opening and closing cash balances at the beginning and end of a quarter. Rows 16–20 detail cash that comes in from customers, grants, and tax credits. Rows 22–33 detail cash that flows out for employees and for expenses such as marketing and rent. Row 29 is fringe benefits which are paid on top of salaries.

Net cash flow, the difference between cash in and out, appears on row 35. Revenue is shown on 39–41. Fixed and variable costs are shown on lines 42–48, leading to a calculation of profit or loss on line 49. The elements of a balance sheet, including assets and liabilities, appear on lines 51–60.

Next, consider the columns of the spreadsheet. Column A lists the rows; detailed explanations appear on the far right of the spreadsheet in column V. Constants that drive some of the calculations are in columns B and C. Column D describes what happens in Quarter 0, as the startup is founded. The first 12 quarters (3 years) appear in columns E-P. Annual summaries of revenue and profit (loss) appear in columns R-T.

Here are the most important regions in such a forecasting model for a startup.

1. All constants specifically input as values are highlighted in yellow. All computed values appear on a white background. The formulas doing the computation may be seen by accessing the spreadsheet at https://ronbaecker.com/ethical-tech-startup-guide/tech-corp-forecasting-rubric/.
2. Monitoring cash balance on rows 15 and 36 is critical to ensuring you do not run out of cash.
3. Equity infusions on line 17 must be timed to finance product development and marketing. The model shows 5 tranches totaling $3.1 million. Ideally, more funds would be raised more quickly, allowing much more rapid investment in sales and marketing.

© The Editor(s) (if applicable) and The Author(s), under exclusive license to Springer Nature Switzerland AG 2023
R. Baecker, *Ethical Tech Startup Guide,* Synthesis Lectures on Professionalism and Career Advancement for Scientists and Engineers,
https://doi.org/10.1007/978-3-031-18780-3

4. The numbers of people hired (rows 7–12) and what you are paying them (rows 23–28) is typically the dominant cost component in the early days of a startup. Sweat equity and stock options must be used to reduce the cash drain. This can be seen as small salaries for principals.

5. Row 3 shows the number of B2B partner customers and row 4 the number of B2C customers. The cost to each B2B customer is set at $500 per quarter, the cost to each B2C customer at $50 per quarter. One row can be eliminated if you do not plan both kinds of business models. It is very difficult to project how quickly you can add customers after your product launch, so this is one of the most fragile parts of the model. In the model shown on the following pages, we assume that the product has been built with sweat equity prior to product launch.

6. The model shows relatively light sales and marketing expenses until quarter 7. Only in Q4 is there a full-time business development person, joined by a second person in Q7. Other marketing expenditures also start to grow significantly in Q7 after the first substantial infusion of investment ($800 K).

7. The model assumes that providing good support is part of the DNA of the company and is a source of competitive advantage. Premium support for B2C customers is priced at $75 per quarter. Market considerations need to be used to set product and support prices (Chap. 5). Design/support/training costs are estimated by assuming an occasional contract designer, and support personnel at 0.1 person per B2B customer and 0.002 person per B2C customer. These numbers will have the be adjusted based on experience.

Note that the model does not include a calculation for taxes (the firm does not achieve profitability until the fourth year) but there is an allowance for the Canadian government reimbursing part of the firm's R&D costs under the SR&ED program (row 19).

Forecasting models are not static entities. They must be tuned month by month to reflect a growing understanding of the financial dynamics of the business. To best understand the model, as you adapt it to your startup and using it as a management tool, you may want to try some of the following exercises:

- Simplify the model so it just represents a startup that is purely B2C with some premium support revenue.
- Expand the model so that the firm sells turnkey systems, which you can do by adding in a new expense line for hardware to be shipped.
- Add a training component of the business, which will require new personnel rows and a revenue row.

Also, try using the model to answer typical questions for which such models are useful. Here are three good questions:

- You discover at the end of Q0 that development has encountered problems and will be two quarters late in producing a usable product. Modify the model, and especially the revenue forecasts, to predict how much additional investment you need and how soon you will need it.
- Although you are concerned that adding more developers will make the late project even later, your VP Development believes that adding two consultants that she knows can reduce the delay in product ship date to one month. Modify the model to predict how much additional investment you need and when.
- By the end of Q3, you realize that you have underestimated the marketing expense required to realize your sales forecasts. Modify the model to predict how much additional investment you need and when.

A	B	C	Q0	Q1	Q2	Q3	Q4	Q5	Q6	Q7	Q8	Q9	Q10
Q0-12 (3 FYs)	2022July26												
SALES ASSUMPTIONS													
B2B partners	$500		2	2	2	3	3	4	5	6	7	9	11
B2C customers	60.0% $50		0	0	10	40	100	300	750	1,200	1,920	3,072	4,915
PERSONNEL ASSUMPTIONS													
CEO			0.2	0.2	0.2	0.2	0.2	0.2	0.4	0.4	0.4	0.7	0.7
VP Development			0.2	0.2	0.2	0.2	0.2	0.2	0.4	0.4	0.4	0.7	0.7
Software developers. #s			0.0	0.0	0.5	1.0	1.0	1.5	2.0	2.0	2.0	2.0	2.0
Biz dev, sales/mktg. #s			0.0	0.2	0.6	0.6	1.0	1.0	1.0	2.0	2.0	2.0	2.5
Design/support/train	0.1	0.002	0.8	0.2	0.2	0.2	0.5	1.5	2.8	3.7	5.2	7.6	11.4
Admin		0.15	0.0	0.0	0.0	0.0	0.0	0.5	1.0	1.3	1.5	1.9	2.6
CASH FLOW													
Opening Balance			$0	$31,020	$13,275	$77,193	$27,803	$111,093	$66,367	$758,546	$583,424	$430,481	$354,910
CASH IN													
Investment			$50,000		$100,000		$150,000		$800,000				
Grants													
SR&Eds	0.35		$0	$0	$0	$0	$0	$33,447	$0	$0	$0	$81,091	$0
Collections	90%			$900	$1,485	$3,690	$7,200	$19,350	$46,125	$72,900	$115,470	$183,762	$292,489
Total Cash In			$50,000	$900	$101,485	$3,690	$157,200	$52,797	$846,125	$72,900	$115,470	$264,853	$292,489
CASH OUT													
CEO	$100,000		$5,000	$5,000	$5,000	$5,000	$5,000	$5,000	$10,000	$10,000	$10,000	$17,500	$17,500
VP Development	$100,000		$5,000	$5,000	$5,000	$5,000	$5,000	$5,000	$10,000	$10,000	$10,000	$17,500	$17,500
S'ware developers	$75,000		$0	$0	$9,375	$18,750	$18,750	$28,125	$37,500	$37,500	$37,500	$37,500	$37,500
Biz dev, sales/mktg	$75,000		$0	$3,750	$11,250	$11,250	$18,750	$18,750	$18,750	$37,500	$37,500	$37,500	$46,875
Design/support/train	$40,000		$7,500	$2,000	$2,000	$2,000	$5,000	$15,000	$27,500	$37,000	$51,900	$75,940	$113,804
Admin	$40,000		$0	$0	$0	$0	$0	$5,000	$9,825	$12,750	$14,985	$19,491	$25,921
Benefits		0.10	$0	$1,575	$3,263	$4,200	$5,250	$7,688	$11,358	$14,475	$16,189	$20,543	$25,910
Legal/accounting			$1,000	$1,000	$1,000	$1,000	$5,000	$1,000	$1,000	$30,000	$5,000	$1,000	$60,000
Rent + office expens	$500		$0	$0	$0	$0	$0	$0	$0	$4,888	$5,744	$7,472	$9,936
Sales/mktg web,sho	30.0%		$0	$0	$0	$5,000	$10,000	$10,000	$25,000	$50,000	$75,000	$100,000	$150,000
Equipment: server, l	$400		$480	$320	$680	$880	$1,160	$1,960	$3,013	$3,910	$4,595	$5,977	$7,949
Total Cash Out			$18,980	$18,645	$37,568	$53,080	$73,910	$97,523	$153,946	$248,023	$268,413	$340,423	$512,895
Cash Flow			$31,020	($17,745)	$63,918	($49,390)	$83,290	($44,726)	$692,180	($175,123)	($152,943)	($75,570)	($220,406)
Closing Balance			$31,020	$13,275	$77,193	$27,803	$111,093	$66,367	$758,546	$583,424	$430,481	$354,910	$134,505
PROFIT AND LOSS													
Product revenue				$1,000	$1,500	$3,500	$6,500	$17,000	$40,000	$63,000	$99,500	$158,100	$251,260
Premium support rev	20%	$75		$0	$150	$600	$1,500	$4,500	$11,250	$18,000	$28,800	$46,080	$73,728
TOTAL REVENUE				$1,000	$1,650	$4,100	$8,000	$21,500	$51,250	$81,000	$128,300	$204,180	$324,988
EXPENSES													
Variable Costs (Cost of Goods Sold)			$7,500	$2,000	$2,000	$2,000	$5,000	$15,000	$27,500	$37,000	$51,900	$75,940	$113,804
Gross Margin			($7,500)	($1,000)	($350)	$2,100	$3,000	$6,500	$23,750	$44,000	$76,400	$128,240	$211,184
Fixed costs			$11,000	$16,325	$34,888	$50,200	$67,750	$80,563	$123,433	$207,113	$211,918	$258,506	$391,142
Hardware depreciation			$40	$67	$123	$197	$293	$457	$708	$1,034	$1,417	$1,915	$2,577
Total Fixed Costs			$11,040	$16,392	$35,011	$50,397	$68,043	$81,019	$124,140	$208,146	$213,334	$260,420	$393,719
TOTAL EXPENSES			$18,540	$18,392	$37,011	$52,397	$73,043	$96,019	$151,640	$245,146	$265,234	$336,360	$507,523
PROFIT (LOSS)			#######	($17,392)	($35,361)	($48,297)	($65,043)	($74,519)	($100,390)	($164,146)	($136,934)	($132,180)	($182,535)
BALANCE SHEET													
ASSETS													
Cash			$31,020	$13,275	$77,193	$27,803	$111,093	$66,367	$758,546	$583,424	$430,481	$354,910	$134,505
Accounts receivable				$100	$165	$410	$800	$2,150	$5,125	$8,100	$12,830	$20,418	$32,499
Equipment			$0	$253	$810	$1,493	$2,360	$3,863	$6,169	$9,045	$12,224	$16,286	$21,658
TOTAL ASSETS			$31,020	$13,628	$78,168	$29,706	$114,253	$72,380	$769,840	$600,569	$455,535	$391,615	$188,662
LIABILITIES													
Accounts payable			$0	$0	$0	$0	$0	$0	$0	$0	$0	$0	$0
TOTAL LIABILITIES			$0	$0	$0	$0	$0	$0	$0	$0	$0	$0	$0
SHAREHOLDER'S EQUITY			$31,020	$13,628	$78,168	$29,706	$114,253	$72,380	$769,840	$600,569	$455,535	$391,615	$188,662

Q9	Q10	Q11	Q12	FY1	FY2	FY3	Notes
							FY0 is 1 quarter only
							Cells in yellow have specific values input, others computed as outputs
9	11	15	20				Specific values for number partners, each yields $C$3 revenue per quarter (Q)
3,072	4,915	7,864	12,583				Specific vals of B2C customers ==> $C$4 rev per Q, then growth of $B$4 per Q
							Salaries for each job category shown in $B$25:$B$30
0.7	0.7	0.7	0.7				#FTE people being paid, reduced fractions for sweat equioty for CEO, VP Dev
0.7	0.7	0.7	0.7				
2.0	2.0	2.5	3.0				
2.0	2.5	3.0	3.0				
7.6	11.4	17.5	27.2				Support people = $B$13*B2B partners + $C$13*B2C customers
1.9	2.6	3.7	5.2				$C$14 admin staff pre core employee (lines 9-13)
$430,481	$354,910	$134,505	$1,990,658				
	$2,000,000						Equity infusions
$81,091	$0	$0	$0				Estimate of SR&Eds based on tech staff, $B$21 of development cost incl.CEO
$183,762	$292,489	$466,813	$745,100				Delay eceiving payments, $B$21 * Rev current Q + (1-$B$21) * Rev previous Q
$264,853	$292,489	$2,466,813	$745,100				Assume no bad debts
$17,500	$17,500	$17,500	$17,500				see line 9
$17,500	$17,500	$17,500	$17,500				see line 10
$37,500	$37,500	$46,875	$56,250				see line 11
$37,500	$46,875	$56,250	$56,250				see line 12
$75,940	$113,804	$174,786	$271,658				see line 13
$19,491	$25,921	$36,568	$51,849				see line 14
$20,543	$25,910	$34,948	$47,101				$C$31 percentage of salary cost for benefits
$1,000	$60,000	$1,000	$5,000				Input specific values for legal and accounting, assume 3 patent filings
$7,472	$9,936	$14,018	$19,875				First no office space, then rent proportional to head count, $B$33 per person
$100,000	$150,000	$200,000	$300,000				Input specific values for sales and marketing expenses
$5,977	$7,949	$11,214	$15,900				$B$34 per employee per quarter
$340,423	$512,895	$610,659	$858,883				
($75,570)	($220,406)	$1,856,154	($113,783)				Cash drain or gain
$354,910	$134,505	$1,990,658	$1,876,875				Note minima in Q1, Q3, Q5, Q10
$158,100	$251,260	$400,716	$639,146				Revenue from B2B and B2C customers
$46,080	$73,728	$117,965	$188,744				A fraction $B$5 of the B2C customers yiled $C$5 revenue per quarter
$204,180	$324,988	$518,681	$827,889	$14,750	$282,050	$1,875,738	TOTAL REVENUE
$75,940	$113,804	$174,786	$271,658				
$128,240	$211,184	$343,894	$556,231				
$258,506	$391,142	$424,659	$571,325				
$1,915	$2,577	$3,512	$4,837				Straight line over 3 years
$260,420	$393,719	$428,170	$576,161				
$336,360	$507,523	$602,957	$847,820				
($132,180)	($182,535)	($84,276)	($19,930)	($166,093)	($475,990)	($418,921)	PROFIT (LOSS). Profitable by Q12
							ASSETS = LIABILITIES + SHAREHOLDER'S EQUITY
$354,910	$134,505	$1,990,658	$1,876,875				
$20,418	$32,499	$51,868	$82,789				Not all revenues are immediately collected
$16,286	$21,658	$29,361	$40,425				Undepreciated value of equipment and furniture
$391,615	$188,662	$2,071,888	$2,000,089				
$0	$0	$0	$0				Assume worst case: all bills paid immediately (as with payroll)
$0	$0	$0	$0				
$391,615	$188,662	$2,071,888	$2,000,089				
				FY1	FY2	FY3	
				2022-23	2023-24	2024-25	

Further Readings

Serial entrepreneur and Stanford Graduate School of Business Profess Steven Brandt presents good advice in *10 Commandments for Building a Growth Company* (Archipelago, 3rd Edition, 1997). His text overlaps with my book in emphasizing qualities of the team and the need to focus, but shows it age with its discussion of traditional business plans and a thin, outdated coverage of financial management and financing. *Business Plans that Win $: Lessons from the MIT Enterprise Forum*, by Stanley Rich and David Gumpert (Harper & Row, 1985), is similar but weaker. Both lack grounding in the spectrum of successful digital technology startups.

Three books early in this century still have value. John Nesheim's *High Tech Start Up: The Complete Handbook for Creating Successful New High Tech Ventures* (The Free Press, 2nd Edition, 2000) focuses primarily on the legal, financial management, and financing aspects of launching a new venture. The strength of Michael Cusomano's *The Business of Software* (Free Press, 2004) is his discussion of running software service and product development firms, but its treatment of markets, value propositions, business models, finances, financing, and team are thin. One-time Apple evangelist and venture capitalist Guy Kawaski's *The Art of the Start: The Time-Tested, Battle-Hardened Guide for Anyone Starting Anything* (Portfolio, 2004) is a lively introduction to the same territory I cover, with special overlap in the areas of positioning, pitching, partnering, and branding, but omits critical topics that I emphasize, most notably related to intellectual property, finances, leadership, management, and team.

Jessica Livingston's *Founders at Work: Stories of Early Startup Days* (APress, 2008) is a series of interviews with the founders of 33 tech startups. The conversations are fascinating, but her questions do not drill deeply enough to elicit comprehensive wisdom. Four of her 33 cases are among my example ventures. *Intentional Integrity: How Smart Companies Can Lead an Ethical Revolution* by Robert Chestnut (Macmillan, 2020), by the Chief Ethics Officer of Airbnb, is a thoughtful discussion of issues important to ethical entrepreneurs, but that is its only topic. It is especially interesting in the light of my discussion of ethical challenges being faced by Airbnb.

© The Editor(s) (if applicable) and The Author(s), under exclusive license to Springer Nature Switzerland AG 2023
R. Baecker, *Ethical Tech Startup Guide,* Synthesis Lectures on Professionalism and Career Advancement for Scientists and Engineers, https://doi.org/10.1007/978-3-031-18780-3

All these books, and mine, are written for novice or would-be entrepreneurs. Three landmark books are aimed at advanced readers. Geoffrey Moore's *Crossing the Chasm: Marketing and Selling Technology Products to Mainstream Customers* (Harper Business, 1991) argues persuasively for the selection of a focused narrow market niche. Clayton Christensen's *The Innovator's Dilemma: The Revolutionary Book That Will Change the Way You Do Business* (Harper Business Essentials, 2000) develops the idea that successful firms need to innovate with disruptive technologies that may supplant their existing product line but fuel the next wave of growth. Eric Ries's *The Lean Startup: How Today's Entrepreneurs Use Continuous Innovation to Create Radically Successful Businesses* (Crown Business, 2011) proposes a build-measure-learn cycle of validated learning to drive product development, an idea which I develop in stronger form in Chap. 3.

There are two relatively current books that are excellent. *Business Model Generation: A Handbook for Visionaries, Game Changers, and Challengers* by Alexander Osterwalder and Yves Pigneur (self-published, businessmodelgeneration.com, 2010) invented the current concept of the business model and business model canvas (discussed in Chap. 5), and is very strong on concepts such as strategy design, user experience, and the go-to-market plan. Wendy Kennedy's *So What? Who Cates? Why You? A Methodology to Find the Business Value of New Ideas* (wendykennedy.com, 2nd Edition, 2015) is strong on problems, solutions, customers, competitive edge, and value proposition. Yet both these books fail to cover issues of intellectual property, competition, financial management, raising finances, management, and leadership.

This book is intended to be a concise introduction, and therefore omits some important topics: non-profit organizations, open source software, and open access publishing. Sources may be found in the Further Reading section of the Bibliography.

Bibliography

Introduction

Bick, Julie. 2005. "The Microsoft Millionaires Come of Age." *The New York Times*. May 29. https://www.nytimes.com/2005/05/29/business/yourmoney/the-microsoft-millionaires-come-of-age.html.

List of Unicorn Startup Companies. 2022. *Wikipedia*. May 30. https://en.wikipedia.org/wiki/List_of_unicorn_startup_companies.

Technology Definition & Meaning. n.d. *Merriam-Webster*. https://www.merriam-webster.com/dictionary/technology.

Technology. n.d. *Encyclopædia Britannica*. https://www.britannica.com/technology/technology.

Chapter 1

Blackberry Limited. 2022. *Wikipedia*. April 21. https://en.wikipedia.org/wiki/BlackBerry_Limited.

Christensen, Clayton M. 2013. *The Innovator's Dilemma: When New Technologies Cause Great Firms to Fail*. Boston, Massachusetts: Harvard Business Review Press.

Christensen, Clayton M., Scott D. Anthony, and Erik A. Roth. 2004. *Seeing What's Next? Using the Theories of Innovation to Predict Industry Change*. Prince Frederick, Maryland: Recorded Books.

Covin, Jeffrey G., Donald F. Kuratko, and Michael H. Morris. 2011. *Corporate Entrepreneurship & Innovation*. Third. Mason, Ohio: South-Western.

Desouza, Kevin C. 2017. *Intrapreneurship: Managing Ideas within Your Organization*. Toronto, Ontario: University of Toronto Press.

Drucker, Peter F. 2006. *Innovation and Entrepreneurship*. Harper Business.

HCR Corporation. 2021. *Wikipedia*. November 8. https://en.wikipedia.org/wiki/HCR_Corporation.

Moore, Geoffrey A. 1998. *Crossing the Chasm*. Chichester: Capstone.

Moore, Geoffrey A. 1999. *Inside the Tornado: Marketing Strategies from Silicon Valley's Cutting Edge*. New York: HarperPerennial.

Nguyen, Clinton. 2016. "7 Cutting-Edge Tech Products That Were Too Early for Their Time." *Business Insider*. July 21. https://www.businessinsider.com/tech-products-that-were-too-early-2016-7.

© The Editor(s) (if applicable) and The Author(s), under exclusive license to Springer Nature Switzerland AG 2023

R. Baecker, *Ethical Tech Startup Guide,* Synthesis Lectures on Professionalism and Career Advancement for Scientists and Engineers, https://doi.org/10.1007/978-3-031-18780-3

Software Arts. 2021. *Wikipedia*. September 10. https://en.wikipedia.org/wiki/Software_Arts.

Vigdor, Neil. 2022. "BlackBerry Ends Service on Its Once-Ubiquitous Mobile Devices." *The New York Times*. January 3. https://www.nytimes.com/2022/01/03/technology/personaltech/black-berry-devices-stop-working.html.

VisiCorp. 2022. *Wikipedia*. April 18. https://en.wikipedia.org/wiki/VisiCorp.

Chapter 2

Boeing 737 MAX. 2022. *Wikipedia*. May 28. https://en.wikipedia.org/wiki/Boeing_737_MAX.

Cagan, Marty. 2018. *Inspired: How to Create Tech Products Customers Love*. Hoboken, New Jersey: John Wiley & Sons, Inc.

Chesbrough, Henry William. 2010. *Open Business Models: How to Thrive in the New Innovation Landscape*. Boston, MA: Harvard Business School Press.

Copyright. 2022. *Wikipedia*. March 28. https://en.wikipedia.org/wiki/Copyright.

Cusumano, Michael A. 2004. *The Business of Software: What Every Manager, Programmer and Entrepreneur Must Know to Thrive and Survive in Good Times and Bad*. New York: Free Press.

Leveson, Nancy G., and Clark S. Turner, "An investigation of the Therac-25 accidents," in Computer, vol. 26, no. 7, pp. 18-41, July 1993, doi: https://doi.org/10.1109/MC.1993.274940.

Millien, Raymond. 2021. "Seven Years after Alice, 63.2% of the U.S. Patents Issued in 2020 Were Software-Related." *IPWatchdog Patents & Patent Law*. March 19. https://www.ipwatchdog.com/2021/03/17/seven-years-after-alice-63-2-of-the-u-s-patents-issued-in-2020-were-software-related/id=130978/.

Mullin, Joe. 2016. "Apple Will Pay $25m to Patent Troll to Avoid East Texas Trial." *Ars Technica*. July 12. https://arstechnica.com/tech-policy/2016/07/apple-will-pay-25m-to-patent-troll-to-avoid-east-texas-trial/.

Mirror Worlds. 2022. *Wikipedia*. February 12. https://en.wikipedia.org/wiki/Mirror_Worlds.

NTP, Inc. 2022. *Wikipedia*. February 17. https://en.wikipedia.org/wiki/NTP,_Inc.#RIM_patent_infringement_litigation.

Patent. 2022. *Wikipedia*. March 14. https://en.wikipedia.org/wiki/Patent.

Trade Secret. 2022. *Wikipedia*. May 29. https://en.wikipedia.org/wiki/Trade_secret.

"Two New Category Pirates Business Books Hit #1 at the Same Time." 2021. *PR.com*. December 7. https://www.pr.com/press-release/850788.

Chapter 3

Donoghue, Karen, and Craig Newell. 2021. *Envision Product: User Experience for Founders*. HumanLogic.

Infinite Renewals: SaaS B2B Onboarding and Professional Services Experts. n.d. *Infinite Renewals*. https://infiniterenewals.com/.

Lewrick, Michael, Patrick Link, and Larry Leifer. 2020. *The Design Thinking Toolbox: A Guide to Mastering the Most Popular and Valuable Innovation Methods*. Hoboken, New Jersey: Wiley.

Market Research. 2022. *Wikipedia*. March 15. https://en.wikipedia.org/wiki/Market_research.

Norman, Donald A. 2013. *The Design of Everyday Things*. New York: Basic Books.

Preece, Jenny, Yvonne Rogers, and Helen Sharp. 2018. *Interaction Design beyond Human-Computer Interaction*. Chichester: Wiley.

Ries, Eric. 2017. *The Lean Startup: How Today's Entrepreneurs Use Continuous Innovation to Create Radically Successful Businesses*. New York: Currency.

Twin, Alexandra. 2021. Market Research Definition. *Investopedia*. September 16. https://www.investopedia.com/terms/m/market-research.asp.

Chapter 4

Crafting an Elevator Pitch: Introducing Your Company Quickly and Compellingly. n.d. *Communications Skills from MindTools.com*. https://www.mindtools.com/pages/article/elevator-pitch.htm.

Kawasaki, Guy. 2015. *The Art of the Start 2.0: The Time-Tested, Battle-Hardened Guide for Anyone Starting Anything*. London, UK: Portfolio Penguin.

McKenna, Regis. 1985. *The Regis Touch: Million-Dollar Advice from America's Top Marketing Consultant*. Addison-Wesley.

Chapter 5

Business Model Canvas. 2022. *Wikipedia*. May 2. https://en.wikipedia.org/wiki/Business_Model_Canvas.

Davidow, William H. 2012. *Marketing High Technology: An Insider's View*. New York: Free Press.

Geyser, Werner. 2022. "What Is an Influencer? - Social Media Influencers Defined [Updated 2022]." *Influencer Marketing Hub*. April 4. https://influencermarketinghub.com/what-is-an-influencer/.

Halligan, Brian. 2022. Inbound Marketing vs. Outbound Marketing. *HubSpot Blog*. April 20. https://blog.hubspot.com/blog/tabid/6307/bid/2989/inbound-marketing-v.

Magretta, Joan. 2002. "Why Business Models Matter." *Harvard Business Review*. May. https://hbr.org/2002/05/why-business-models-matter.

Robertson-Adams, Grant. 2022. "Virality Explained: 6 Ways to Design a Viral Product." *GrowSurf*. Accessed May 31. https://growsurf.com/blog/virality-explained-viral-design.

Search Engine Optimization. 2022. *Wikipedia*. April 13. https://en.wikipedia.org/wiki/Search_engine_optimization.

Singh, Shiv, and Stephanie Diamond. 2020. *Social Media Marketing for Dummies*. Hoboken, New Jersey: J. Wiley & Sons.

Statista Research Department. 2022a. "Time Spent with Digital vs. Traditional Media in the U.S. 2023." *Statista*. May 3. https://www.statista.com/statistics/565628/time-spent-digital-traditional-media-usa/.

Statista Research Department. 2022b. "Most Used Social Media 2021." *Statista*. March 8. https://www.statista.com/statistics/272014/global-social-networks-ranked-by-number-of-users/.

Twin, Alexandra. 2022. "The 4 PS of Marketing." *Investopedia*. March 2. https://www.investopedia.com/terms/f/four-ps.asp.

Virality: Master Product Virality, Benefits of Virality, How Virality Works. 2021. *Platform Thinking Labs*. July 31. https://platformthinkinglabs.com/materials/virality-viral-growth-network-effects/.

Chapter 6

Bankruptcy. 2022. *Wikipedia*. May 14. https://en.wikipedia.org/wiki/Bankruptcy.

FTX, 2022, *Wikipedia*. Dec. 15. https://en.wikipedia.org/wiki/FTX_(company).

Osnabrugge, Van Mark, and Robert J. Robinson. 2000. *Angel Investing: Matching Startup Funds with Startup Companies: The Guide for Entrepreneurs, Individual Investors, and Venture Capitalists*. San Francisco, California: Jossey-Bass.

Rich, Stanley R., and David E. Gumpert. 1985. *Business Plans That Win Three Dollar Signs: Lessons from the MIT Enterprise Forum*. New York: Harper & Row.

Theranos. 2022. *Wikipedia*. May 26. https://en.wikipedia.org/wiki/Theranos.

Woo, Erin, and Maureen Farrell. 2022. "Bolt Built $11 Billion Payment Business on Inflated Metrics and Eager Investors." *The New York Times*. May 10. https://www.nytimes.com/2022/05/10/business/bolt-start-up-ryan-breslow-investors.html.

Chapter 7

Bhattacharjee, Rajarshi. 2012. "Culture Is What People Do When No One Is Looking: Gerard Seijts." *Business Standard*. September 23. https://www.business-standard.com/article/management/culture-is-what-people-do-when-no-one-is-looking-gerard-seijts-112092400046_1.html.

Coutu, Sherry. 2014. "The Scale-up Report on UK Economic Growth Sherry Coutu CBEǀ - LSE Home." *LSE Entrepreneurship*. November. https://www.lse.ac.uk/assets/richmedia/channels/publicLecturesAndEvents/slides/20141118_1800_scaleupManifesto_sl.pdf.

Grove, Andrew S. 2002. *Only the Paranoid Survive: How to Exploit the Crisis Points That Challenge Every Company and Career*. London, UK: Profile Books.

How to Build a Startup Culture That Everybody Wants. 2021. *Visible.vc*. July 30. https://visible.vc/blog/startup-culture/.

Kim, W. Chan, and Renée Mauborgne. 2020. *Blue Ocean Shift - Beyond Competing: Proven Steps to Inspire Confidence and Seize New Growth*. London, UK: Macmillan.

Leadership. 2022. *Wikipedia*. May 27. https://en.wikipedia.org/wiki/Leadership.

Legault, Dan. 2021a. Why You Should Use Command-Post Management If a Pivot Is in Your near Future. *Inc.com*. November 28. https://www.inc.com/dan-legault/why-you-should-use-command-post-management-if-a-pivot-is-in-your-near-future.html.

Mar, Anna. 2014. "7 Definitions of Leadership." *Simplicable*. May 18. https://training.simplicable.com/training/new/7-definitions-of-leadership.

McAfee, Sam. 2017. "Patterns of Good Startup Culture (Ch. 13)." *Medium*. May 31. https://medium.com/startup-patterns/patterns-of-good-startup-culture-ch-13-dfba4323ec8e.

Mintzberg, Henry. 1989. *Mintzberg on Management: Inside Our Strange World of Organizations*. New York: Free Press.

Tank, Aytekin. 2017. "How to Build a Great Startup Culture." *Entrepreneur*. Entrepreneur. February 20. https://www.entrepreneur.com/article/287927.

Tekir, Arzu. 2020. "Culture Matters: How Great Startups Will Thrive in 2020." *Forbes*. February 21. https://www.forbes.com/sites/ellevate/2020/02/11/culture-matters-how-great-startups-will-thrive-in-2020/?sh=25850fb67c76.

Thiel, Peter, and Blake Masters. 2015. *Zero to One: Notes on Startups, or How to Build the Future*. London: Virgin Books.

Wiles, Jackie. 2021. "The 5 Pillars of Strategy Execution." *Gartner*. April 26. https://www.gartner.com/smarterwithgartner/the-five-pillars-of-strategy-execution.

Wyatt, Frank J. 2019. "What Is Business Execution and Why Should I Care about It?" *Medium*. July 11. https://medium.com/business-process-management-software-comparisons/ what-is-business-execution-and-why-should-i-care-about-it-c0b2aebdbbc8.

Chapter 8

Airbnb Code of Ethics. 2020. *Airbnb*. December 13. https://s26.q4cdn.com/656283129/files/doc_ downloads/governance_doc_updated/Code_of_Ethics.pdf.

Chestnut, Robert. 2021. *Intentional Integrity: How Smart Companies Can Lead an Ethical Revolution - and Why That's... Good for All of Us*. Pan Books.

Cooke, Rachel. 2020. "How Can You (and Your Workplace) Embrace Intentional Integrity?" *Quick and Dirty Tips*. August 18. https://www.quickanddirtytips.com/business-career.

Conger, Kate. 2022a. "Inside Twitter, Fears Musk Will Return Platform to Its Early Troubles." *The New York Times*. April 28. https://www.nytimes.com/2022a/04/28/technology/twitter-musk-content-moderators.html.

Editors, BT. 2019. "Ethical Behavior - Definition, Examples and Quiz." *Business Terms*. January 8. https://businessterms.org/ethical-behavior/.

Edwards, Steph. 2021. "Is There an Ethical Problem with Using Airbnb?" *The Mediterranean Traveller*. January 5. https://www.themediterraneantraveller.com/airbnb-ethically/.

Espósito, Filipe. 2021. "Users Spent $133 Billion with Apps in 2021; App Store Has Higher Revenue than Google Play." *9to5Mac*. December 7. https://9to5mac.com/2021/12/07/ users-spend-133-billion-with-apps-in-2.

Ethical. n.d. *Cambridge Dictionary*. https://dictionary.cambridge.org/dictionary/english/ethical.

Le Roux, Jano. 2022. "Apple Just Wrecked 15+ Startups in Less than 1 Hour." *Medium*. The Startup. June 10. https://medium.com/swlh/ apple-just-wrecked-15-startups-in-less-than-1-hour-ca1593b2ca7f.

Levy, Steven. 2022a. "Plaintext: Even Twitter Thinks Elon Musk's Tweets Are out of Control." *Wired*. July 15. https://www.wired.com/story/plaintext-elon-musk-tweets/.

Maxwell, Mackenzie. 2021. "What Is Ethical Behavior in a Workplace?" *Bizfluent*. November 20. https://bizfluent.com/info-8095779-ethical-behavior-workplace.html.

Mickle, Tripp. 2022. "How Technocrats Triumphed at Apple." *The New York Times*. May 1. https:// www.nytimes.com/2022/05/01/technology/jony-ive-apple-design.html.

Microsoft. 2022. "Microsoft Responsible AI Standard, V2." *Microsoft*. June. https://blogs.microsoft.com/wp-content/uploads/prod/sites/5/2022/06/Microsoft-Responsible-AI-Standard-v2-General-Requirements-3.pdf.

Nicas, Jack, and Flávia Milhorance. 2022. "In Brazil, Firms Sought Black Workers. Then Linkedin Got Involved." *The New York Times*. April 2. https://www.nytimes.com/2022/04/02/world/americas/linkedin-brazil-jobs-diversity.html.

Pegoraro, Rob. 2022. "5 Technologies That Should Give Us Some Hope for the Planet's Future." *Fast Company*. April 22. https://www.fastcompany.com/90744356/ five-technologies-that-should-give-us-some-hope-for-the-planets-future.

Seagle, Cameron. 2022. "Airbnb Ethical Issues That Are Cause for Concern." *The World Pursuit*. March 17. https://theworldpursuit.com/airbnb-ethical-issues/.

Slivka, Eric. 2022. "'After Steve' Examines the Tensions That Led to Jony Ive's Departure from Apple." *MacRumors*. May 1. https://www.macrumors.com/2022/05/01/ after-steve-book-jony-ive/.

Smith, Brad. 2018. "Facial Recognition: It's Time for Action." *Microsoft On the Issues*. Microsoft. December 6. https://blogs.microsoft.com/on-the-issues/2018/12/06/facial-recognition-i.

Stiffler, Lisa. 2019. "From 'Evil Empire' to Model Citizen? How Microsoft's Good Deeds Work to Its Competitive Advantage." *GeekWire*. December 18. https://www.geekwire.com/2019/evil-empire-model-citizen-microsofts-good-deeds-work-competitive-advantage/.

Tennant, Don. 2021. "Debunking the Myth of Microsoft as the 'Evil Empire'." *IT Business Edge*. March 10. https://www.itbusinessedge.com/it-management/debunking-the-myth-of.

Vincent, James. 2022. "Microsoft to Retire Controversial Facial Recognition Tool That Claims to Identify Emotion." *The Verge*. June 21. https://www.theverge.com/2022/6/21/23177016/microsoft-retires-emoti.

Wadhwa, Vivek, Ismail Amla, and Alex Salkever. 2021. "How Microsoft Transformed Itself from Evil Empire to Cool Kid." *Fortune*. December 21. https://fortune.com/2021/12/21/microsoft-cultural-transformation-book-excerpt-satya-nadella/.

What Is Ethical Behavior? 2020. *Reference*. IAC Publishing. March 26. https://www.reference.com/world-view/ethical-behavior-df19631993d13e54.

Chapter 9

Bell, C. Gordon, and John E. MacNamara. 2000. *High-Tech Ventures: The Guide for Entrepreneurial Success*. Reading, Massachusetts: Addison-Wesley.

Brandt, Steven C. 1997. *Entrepreneuring: The 10 Commandments for Building a Growth Company*. Friday Harbor, Washington: Archipelago Publications.

Kennedy, Wendy, Peter Eddison, and John Hanak. 2015. *So What? Who Cares? Why You?: A Methodology to Find the Business Value of New Ideas*. Ottawa, Ontario: Wendykennedy.com Inc.

Levering, Robert, Michael Katz, and Milton Moskowitz. 1984. *The Computer Entrepreneurs: Who's Making It Big and How in America's Upstart Industry*. New York: New American Library.

Livingston, Jessica. 2008a. *Founders at Work Stories of Startups: Early Days*. Apress.

Nesheim, John L. 2000. *High Tech Start-Up: The Complete Handbook for Creating Successful New High Tech Companies*. New York: Free Press.

Osterwalder, Alexander, and Yves Pigneur. 2013. *Business Model Generation: A Handbook for Visionaries, Game Changers, and Challengers*. New York: Wiley & Sons.

A. Imax

Acland, Charles. 2012. "IMAX Systems Corporation." *The Canadian Encyclopedia*. December 3. https://www.thecanadianencyclopedia.ca/en/article/imax-systems-corporation.

Barnes, Brooks. 2022. "IMAX Looks beyond Movies to Live Events." *The New York Times*. February 23. https://www.nytimes.com/2022/02/23/business/media/imax-movies-live-events.html.

Disse, Diane. 2022. "The Birth of IMAX." *IEEE*. https://ewh.ieee.org/reg/7/millennium/imax/imax_birth.html.

IMAX Corporation Reports Q4 and Full-Year 2021 Results. 2022. *Yahoo! Finance*. February 23. https://finance.yahoo.com/news/imax-corporation-reports-q4-full-211900239.html.

IMAX. 2022. *Wikipedia*. March 27. https://en.wikipedia.org/wiki/IMAX.

B. Microsoft

Cusumano, Michael A., and Richard W. Selby. 1995. *Microsoft Secrets: How the World's Most Powerful Software Company Creates Technology, Shapes Markets, and Manages People*. New York: Free Press.

Denning, Steve. 2021. "How Microsoft's Digital Transformation Created a Trillion Dollar Gain." *Forbes*. June 30. https://www.forbes.com/sites/stevedenning/2021/06/20/how-microsofts-digital-transformation-created-a-trillion-dollar-gain/?sh=68267bd1625b.

Gates, Bill, and Reid Hoffman. 2019. "MoS Episode Transcript – Bill Gates." *Masters of Scale*. November. https://mastersofscale.com/wp-content/uploads/2019/11/mos-episode-transcript-bill-gates-.pdf.

Manes, Stephen, and Paul Andrews. 1994. *Gates: How Microsoft's Mogul Reinvented an Industry and Made Himself the Richest Man in America*. New York: Touchstone.

Microsoft. 2022. *Wikipedia*. March 28. https://en.wikipedia.org/wiki/Microsoft.

Microsoft Is Great Again' under Satya Nadella. A Shareholder's Dream Stock. 2019. *SeekingAlpha*. January 3. https://seekingalpha.com/instablog/32694515-stockmarketsquawk/5254786-microsoft-is-great-again-under-satya-nadella-shareholder-s-dream-stock.

Stross, Randall E. 1998. *The Microsoft Way: The Real Story of How the Company Outsmarts Its Competition*. London: Warner.

Water, Richard. 2019. "Satya Nadella Brought Microsoft Back from the Brink of Irrelevance." *Los Angeles Times*. December 21. https://www.latimes.com/business/technology/story/2019-12-21/satya-nadella-reinvigorated-microsoft.

C. Apple Inc.

Apple Inc. 2019. *Wikipedia*. October 9. https://en.wikipedia.org/wiki/Apple_Inc.

Carlton, Jim. 1997. *Apple: The inside Story of Intrigue, Egomania, and Business Blunders That Toppled an American Icon*. New York: Time Books/Random House.

Enjeti, Anjali. 2014. "Jonathan Zufi Captures Apple's Evolution in Iconic, an Encyclopedic Picture Book." *Arts Atl*. January 28. https://www.artsatl.org/jonathan-zufi-captures-apples-evolution-iconic-e.

Isaacson, Walter. 2021. *Steve Jobs*. New York: Simon & Schuster Paperbacks.

June, Laura. 2011. "Book Review: 'Steve Jobs' by Walter Isaacson." *The Verge*. October 27. https://www.theverge.com/2011/10/27/2517152/book-review-steve-jobs.

Karney, Leander. 2008. "How Apple Got Everything Right by Doing Everything Wrong - Wired." *Wired*. March 18. https://www.wired.com/2008/03/how-apple-got-everything-right-by-doing-everything-wrong/.

Malone, Michael S. 2000. *Infinite Loop: How Apple, the World's Most Insanely Great Computer Company, Went Insane*. London, UK: Aurum.

West Coast Computer Faire. 2022. *Wikipedia*. March 14. https://en.wikipedia.org/wiki/West_Coast_Computer_Faire.

D. Adobe Inc.

Adobe Inc. 2018. *Wikipedia*. October 12. https://en.wikipedia.org/wiki/Adobe_Inc.

Knowledge at Wharton Staff. 2010. "Adobe Co-Founder John Warnock on the Competitive Advantages of Aesthetics and the 'Right' Technology." *Knowledge at Wharton*. January 10. https://knowledge.wharton.upenn.edu/article/adobe-co-founder-john-wa.

Knowledge at Wharton Staff. 2008. "Driving Adobe: Co-Founder Charles Geschke on Challenges, Change and Values." *Knowledge at Wharton*. September 3. https://knowledge.wharton.upenn.edu/article/driving-adobe-co-founder-charles-geschke-on-challenges-change-and-values/.

E. SideFX

About SideFX. 2022. *SideFx*. https://www.sidefx.com/company/about-sidefx/.

Kim Davidson and Greg Hermanovic: Graphics Interface. 2017. *Graphics Interface*. September 16. https://graphicsinterface.org/awards/cdmp/davidson-hermanovic/.

Seymour, Mike. 2012. Side Effects Software – 25 Years On. *Fxguide*. February 27. https://www.fxguide.com/fxfeatured/side-effects-software-25-years-on/.

F. Caseware International

Brunner, Ryan. 2019. "CaseWare International Interview Questions." *CaseWare International Interview Questions*. December 27. https://www.mockquestions.com/company/CaseWare+International%2C+Inc./.

"CaseWare CA: Cloud-Based Audit & Accounting Software." *CaseWare*. https://caseware.com/ca.

Caseware International. 2022. *Wikipedia*. March 14. https://en.wikipedia.org/wiki/CaseWare_International.

President & CEO of CaseWare International Inc. Designated Fellow by CPA Ontario. 2017. *CaseWare*. December 5. https://www.caseware.com/ca/blog/dwight-wainman-president-ceo-caseware-international-inc-designated-fellow-cpa-ontario#:~:text=CaseWare%20International%20Inc.%20is%20proud,profession%20and%20designated%20a%20Fellow.

G. New York Times Reinvented

Decline of Newspapers. *Wikipedia*, April 30, 2022. https://en.wikipedia.org/wiki/Decline_of_newspapers.

Doctor, Ken. 2020. "Newsonomics: The New York Times' New CEO, Meredith Levien, on Building a World-Class Digital Media Business - and a Tech Company." *Nieman Lab*. July 30. https://www.niemanlab.org/2020/07/newsonomics-the-new-york-times-new-ceo-meredith-levien-on-building-a-world-class-digital-media-business-and-a-tech-company/.

Meredith Kopit Levien. *Wikipedia*, May 13, 2022. https://en.wikipedia.org/wiki/Meredith_Kopit_Levien.

The New York Times. *Wikipedia*, May 24, 2022. https://en.wikipedia.org/wiki/The_New_York_Times.

Spears, Jared. "The New York Times 'Subscriber-First' Strategy - State of Digital Publisher." *State of Digital Publishing*, January 22, 2019. https://www.stateofdigitalpublishing.com/opinion/pursuing-subscriber-first-strategy-york-times-learns-consumer-brand/.

SR2020. "(Turn) off the Press: The New York Times Is Winning with Digital." *Digital Innovation and Transformation*, February 11, 2020. https://digital.hbs.edu/platform-digit/submission/turn-off-the-press-the-new-york-times-is-winning-with-digital/.

H. Shaw Industries Reinvented

Circular Economy. 2022. *Wikipedia*. May 29. https://en.wikipedia.org/wiki/Circular_economy.

Cradle to Cradle Impact Study Technical Report. *Cradle to Cradle Products Innovation Institute*, n.d. https://cdn.c2ccertified.org/resources/impact_study_technical_report.pdf.

History. n.d. *Shaw Industries Group, Inc.* https://shawinc.com/Company-Profile/History.

Shaw's ECOWORX Carpet Tile Deemed 'Cradle to Cradle' Certified. *Floor Covering News*, October 4, 2021. https://www.fcnews.net/2021/10/shaws-ecoworx-carpet-tile-deemed-crad.

Shaw Industries Simultaneously Focuses on Environmental Health and the Human Experience. 2019. *Floor Trends Magazine*. August 8. https://www.floortrendsmag.com/articles/105170-shaw-industries-simultaneously-focuses-on-environmental-health-and-the-human-experience.

Shaw Industries Simultaneously Focuses on Environmental Health and the Human Experience. n.d. *Shaw Industries Group, Inc.* https://shawinc.com/Newsroom/Press-Releases/Shaw-Industries-Simultaneously-Focuses-on-Environm.

Sustainability Report 2021. *Shaw Incorporated*, 2021. https://shawinc.com/2021sustainabilityreport.

I. Desire2Learn (D2L)

Carter, Dennis. 2009. "Blackboard, D2L Declare Legal Truce." *ESchool News*. December 17. https://www.eschoolnews.com/2009/12/16/blackboard-d2l-declare-legal-truce/.

D2L. 2022. *Wikipedia*. February 26. https://en.wikipedia.org/wiki/D2L.

EPC Spotlights... Desire2Learn. 2008. *The Wayback Machine*. Texas Distance Learning Association . July 18. https://web.archive.org/web/20080828054609/http://txdla.org/documents/EPCspotlight/desire2learn.htm.

Fenton, William. 2017. "D2L Brightspace LMS Review." *PCMAG*. October 5. https://www.pcmag.com/reviews/d2l-brightspace-lms.

Hill, Phil. 2018. "University of Wisconsin System to Migrate from D2L Brightspace to Canvas LMS." *ELiterate*. January 3. https://eliterate.us/university-wisconsin-system-migrate-d2l-brightspace-canvas-lms/.

Rogers, Bruce. 2014. "John Baker's Desire2Learn Is Leading Education Transformation." *Forbes*. April 24. https://www.forbes.com/sites/brucerogers/2014/04/23/john-bakers-desire2learn-is-leading-education-transformation/?sh=429e6f613ff0.

J. Wikipedia

History of Wikipedia. 2022. *Wikipedia*. April 1. https://en.wikipedia.org/wiki/History_of_Wikipedia.
Jimmy Wales. 2022. *Wikipedia*. May 28. https://en.wikipedia.org/wiki/Jimmy_Wales.
Wikipedia. 2022. *Wikipedia*. April 4. https://en.wikipedia.org/wiki/Wikipedia.
Wikipedia: About. 2022. *Wikipedia*. March 31. https://en.wikipedia.org/wiki/Wikipedia:About.
Wikipedia's Founder on How the Site Was Built & Promoted - with Jimmy Wales. 2009. *Business Podcast for Startups*. July 7. https://mixergy.com/interviews/wikipedias-founder-jimmy-wales/.

K. LinkedIn

Bishop, Todd. 2021. "LinkedIn Posts First $10 Billion Year, 5 Years after Microsoft Deal, but Profits Remain a Mystery." *GeekWire*. July 28. https://www.geekwire.com/2021/linkedin-posts-first-10-billion-year-5-years-microsoft-deal-profits-remain-mystery/.
Erlichman, Jon. 2019. "Three Years after Microsoft Acquisition, Linkedin Keeps Quietly Climbing - Bnn Bloomberg." *BNN*. October 23. https://www.bnnbloomberg.ca/three-years-after-microsoft-acquisition-linkedin-keeps-quietly-climbing-1.1335990.
Geekmaster. 2020. "Jeff Weiner - Leadership Style & Principles." *Geeknack*. October 10. https://www.geeknack.com/2020/10/10/jeff-weiner-leadership-style-principles/.
Hempel, Jessi. 2017. "Now We Know Why Microsoft Bought LinkedIn." *Wired*. March 14. https://www.wired.com/2017/03/now-we-know-why-microsoft-bought-linkedin/.
Hoffman, Reid, and Matthew Mercer. 2019. "MoS Episode Transcript: Reid Hoffman." *Masters of Scale*. January. https://mastersofscale.com/wp-content/uploads/2019/01/mos-episode-transcript_-reid-hoffman.pdf.
Hoffman, Reid, and June Cohen. 2022. "MoS Episode Transcript – Reid Hoffman Part 2." *Masters of Scale*. https://mastersofscale.com/wp-content/uploads/2019/02/mos-episode-transcript-reid-hoffman-part-2.pdf.
Hoffman, Reid. 2012. "PandoMonthly: Fireside Chat With Reid Hoffman." *Ghost Archive*. August 11. https://ghostarchive.org/varchive/lKDcbFGct8A.
How Does Linkedin Make Money? 2021. *Investopedia*. June 9. https://www.investopedia.com/ask/answers/120214/how-does-linkedin-lnkd-make-money.asp.
Iqbal, Mansoor. 2022. "LinkedIn Usage and Revenue Statistics (2022)." *Business of Apps*. January 11. https://www.businessofapps.com/data/linkedin-statistics/.
Jeff Weiner. 2022. *Wikipedia*. February 7. https://en.wikipedia.org/wiki/Jeff_Weiner.
LinkedIn. 2022. *Wikipedia*. April 3. https://en.wikipedia.org/wiki/LinkedIn.
Microsoft-LinkedIn Timeline: An inside Look at the Merger. 2021. *Wall Street Prep*. September 28. https://www.wallstreetprep.com/knowledge/a-look-at-the-microsoft-linkedin-merger/.
Reid Hoffman. 2022. *Wikipedia*. March 15. https://en.wikipedia.org/wiki/Reid_Hoffman.

L. Antibe Therapeutics

Antibe Therapeutics (ATE) CEO Dan Legault Discussion, Deep Dive, Q&A - Otenaproxesul Pain Trials. 2021. *YouTube*. https://www.youtube.com/watch?v=AR0wzH5T_Ac.
Antibe Therapeutics -This Toronto Biotech Firm Is Developing Safer Treatments for Pain and Inflammation. 2022. *Innovations Of The World*. March 30. https://innovationsoftheworld.com/antibe-therapeutics-this-toronto-biotech-firm-is-developing-safer-treatments-for-pain-and-inflammation/.

Antibe Therapeutics. 2022. *Wikipedia*. March 1. https://en.wikipedia.org/wiki/Antibe_Therapeutics.

Dan Legault - CEO - Antibe Therapeutics Inc - Linkedin. 2022. *LinkedIn*. https://ca.linkedin.com/in/dan-legault-84125033.

John L. Wallace. 2022. *Wikipedia*. January 28. https://en.wikipedia.org/wiki/John_L._Wallace.

Legault, Dan. 2021b. "Why You Should Use Command-Post Management If a Pivot Is in Your near Future." *Inc*. November 28. https://www.inc.com/dan-legault/why-you-should-use-command-post-management-if-a-pivot-is-in-your-near-future.html.

Our Story: Targeting Large Markets in Pain and Inflammation... 2022. *Antibe Therapeutics*. Accessed April 5. https://www.antibethera.com/about-us/our-story/.

Sekhon, Sandip. 2019. "Global Chronic Pain Statistics [2018 Infographic]." *Pathways*. October 19. https://www.pathways.health/global-chronic-pain-statistics-2018-infographic/.

Wallace, John L., and Rui Wang. 2015. "Hydrogen Sulfide-Based Therapeutics: Exploiting a Unique but Ubiquitous Gasotransmitter." *Nature Reviews Drug Discovery* 14 (5): 329–45. doi:https://doi.org/10.1038/nrd4433.

Why Hydrogen Sulfide? 2022. *Antibe Therapeutics*. https://www.antibethera.com/science/why-h2s/.

M. Twitter

Conger, Kate, and Lauren Hirsch. 2021. "Twitter's Jack Dorsey Steps down from C.E.O. Role." *The New York Times*. November 29. https://www.nytimes.com/2021/11/29/technology/jack-dorsey-twitter.html.

Conger, Kate. 2022. "Twitter Wants to Reinvent Itself, by Merging the Old with the New." *The New York Times*. March 2. https://www.nytimes.com/2022c/03/02/technology/twitter-platform-rethink.html.

Conger, Kate. 2022. "Twitter Expands Content-Moderation Rules to Cover Crises like War and Disasters." *The New York Times*. May 19. https://www.nytimes.com/2022b/05/19/business/twitter-content-moderation.html.

Conger, Kate, Mike Isaac, Ryan Mac, and Tiffany Hsu. 2022. "Two Weeks of Chaos: Inside Elon Musk's Takeover of Twitter". The New York Times. November 11. https://www.nytimes.com/2022/11/11/technology/elon-musk-twitter-takeover.html.

Dorsey, Jack. 2022. *Wikipedia*. March 27. https://en.wikipedia.org/wiki/Jack_Dorsey.

Frenkel, Sheera and Kate Conger, 2022. "Hate Speech's Rise on Twitter Is Unprecedented, Researchers Find". The New York Times. December 2. https://www.nytimes.com/2022/12/02/technology/twitter-hate-speech.html.

Levy, Steven. 2022b. "Can Elon Musk Make Twitter Great Again?" *Wired*. April 8. https://www.wired.com/story/plaintext-elon-musk-twitter-great-again/.

Mac, Ryan, Mike Isaac, and Kellen Browning, 2022. "Elon Musk's Twitter Teeters on the Edge After Another 1,200 Leave". "The New York Times. November 18. https://www.nytimes.com/2022/11/18/technology/elon-musk-twitter-workers-quit.html.

Pariser, Eli, 2022. "Musk's Twitter Will Not Be the Town Square the World Needs". Wired. October 28. https://www.wired.com/story/elon-musk-twitter-town-square/.

Reiff, Nathan. 2022. "How Twitter Makes Money: Advertising Comprises the Bulk of Revenue." *Investopedia*. April 6. https://www.investopedia.com/ask/answers/120114/how-does-twitter-twtr-make-money.asp.

Shenvi, Dipak. 2021. "How Does Twitter Make Money: Analyzing the Revenue Model." *The Strategy Story*. July 4. https://thestrategystory.com/2021/05/21/twitter-revenue-model/.

Swisher, Kara. 2022. "Twitter's Former C.E.O. Has a 'Too Bad, so Sad' Approach to Content Moderation." *The New York Times*. The New York Times. January 3. https://www.nytimes.com/2022/01/03/opinion/sway-dick-costolo-kara-swisher.html.

Twitter by the Numbers (2022): Stats, Demographics & Fun Facts." 2022. *Omnicore*. https://www.omnicoreagency.com/twitter-statistics/.

Twitter. 2022. *Wikipedia*. April 7. https://en.wikipedia.org/wiki/Twitter.

Woo, Erin. 2022. "Twitter Will Stiffen Moderation Policies in Response to the War in Ukraine." *The New York Times*. April 5. https://www.nytimes.com/2022/04/05/business/twitter-policy-ukraine.html.

N. QLess

Bäcker, Alex. 2022. "Who Alex Bäcker Is." *Alex Bäcker's Wiki* . March 28. http://alexbacker.pbworks.com/w/page/1721233/Who%2520Alex%2520B%C3%A4cker%2520is.

Null, Christopher, and Freelance Contributor. 2013. "Mobile Queuing App Promises to Take the Pain out of Waiting in Line." *PCWorld*. July 12. https://www.pcworld.com/article/452862/mobile-queuing-app-promises-to-take-the-pain-out-of-waiting-in-line.html.

QLess. 2022. February 2. https://qless.com/.

O. Ecovative Design

About Ecovative: Our Story. n.d. *Ecovative*. https://ecovative.com/our-story.

Ecovative Design. 2022. *Wikipedia*. February 21. https://en.wikipedia.org/wiki/Ecovative_Design.

Frazier, Ian. 2013. "Form and Fungus." *The New Yorker*. May 13. https://www.newyorker.com/magazine/2013/05/20/form-and-fungus.

Hammond, Dawn. 2020. "IKEA Commits to Biodegradable Mushroom Packaging." *Yahoo! News*. Yahoo! February 4. https://news.yahoo.com/ikea-commits-biodegradable-mushroom-packaging-220023480.html.

Scanlon, Jessie. 2009. "NCIIA: Promoting Student Inventors." *Bloomberg Business*. Web Archive. March 23. https://web.archive.org/web/20150816001300/http://www.businessweek.com/innovate/content/mar2009/id20090323_482784.htm.

P. Chess.com

Allebest, Erik. 2020. "Incredible Second Wave of Interest in Chess." *Chess.com*. November 22. https://www.chess.com/blog/erik/incredible-second-wave-of-interest-in-chess.

Chess.com. 2007a. "Building Chess.com: Part 1 - Getting Started." *Chess.com*. August 19. https://www.chess.com/news/view/building-chesscom-part-1---getting-st.

Chess.com. 2007b. "Building Chess.com: Part 2 - Putting Pieces Together." *Chess.com*. August 10. https://www.chess.com/news/view/building-chesscom-part-2---putting-pieces-together.

Chess.com. 2007c. "Building Chess.com: Part 3 - Growing Pains." *Chess.com*. August 12. https://www.chess.com/news/view/building-chesscom-part-3---growing-pains.

Chess.com. 2016. "The Chess.com Logo Story." *Chess.com*. July 22. https://www.chess.com/blog/erik/the-chess-com-logo-story.

Chess.com. 2019a. "How I Got the Chess.com Domain Name." *Chess.com*. August 2. https://www.chess.com/blog/erik/how-i-got-the-chess-com-domain-name.

Chess.com. 2019b. "Why (and How) I Started Chess.com." *Chess.com*. January 9. https://www.chess.com/blog/erik/why-and-how-i-started-chess-com.

Chess.com. 2021. "How Chess.com's 100-Person Virtual Team Works Together." *Chess.com*. November 8. https://www.chess.com/blog/erik/how-chess-com-s-100-person-virtual-team-works-together.

Chess.com. 2022. *Wikipedia*. March 20. https://en.wikipedia.org/wiki/Chess.com.

Q. Airbnb

Airbnb. 2022. *Wikipedia*. April 4. https://en.wikipedia.org/wiki/Airbnb.

Airbnb Financial Reports. 2020. *U.S. Securities and Exchange Commission*. https://www.sec.gov/Archives/edgar/data/1559720/000119312520294801/d81668ds1.htm.

Airbnb IPO. 2020. *U.S. Securities and Exchange Commission*. November 16. https://www.sec.gov/Archives/edgar/data/1559720/000119312520294801/d81668ds1.htm.

Airbnb. 2021. "What Makes Airbnb, Airbnb." *Airbnb Newsroom*. December 8. https://news.airbnb.com/what-makes-airbnb-airbnb/.

Chesky, Brian, and Reid Hoffman. 2021. "Masters of Scale Episode Transcript: 100th Episode." *Masters of Scale*. November. https://mastersofscale.com/wp-content/uploads/2021/11/masters-of-scale-episode-transcript_-100th-episode.pdf.

Damchevski, Goran. 2022. "Why Airbnb's (Nasdaq:ABNB) High Growth May Be Already Factored-In." *Yahoo! Finance*. March 21. https://ca.finance.yahoo.com/news/why-airbnbs-nasdaq-abnb-high-030127546.html.

McCann, Chris. 2015. "Scaling Airbnb with Brian Chesky - Class 18 Notes of Stanford University's CS183C." *Medium*. Blitzscaling: Class Notes and Essays. December 8. https://medium.com/cs183c-blitzscaling-class-collection/scaling-airbnb-with-brian-chesky-class-18-notes-of-stanford-university-s-cs183c-3fcf75778358.

Seagle, Cameron. 2020. "Brian Chesky on Masters of Scale." *Masters of Scale*. September 4. https://mastersofscale.com/brian-chesky-handcrafted/.

R. Beyond Meat

Beyond Meat. *Wikipedia*, April 6, 2022. https://en.wikipedia.org/wiki/Beyond_Meat.

Bowler, Dane. "Beyond Meat: Unhealthy for the Heart and the Portfolio." *Seeking Alpha*, May 9, 2019. https://seekingalpha.com/article/4261728-beyond-meat-unhealthy-for-heart-and-portfolio.

Ethan Brown. *Beyond Meat, Inc.*, n.d. https://investors.beyondmeat.com/board-member-management/ethan-brown#:~:text=Ethan%20Brown%20is%20the%20founder,since%20our%20inception%20in%202009.

Jacobsen, Rowan. "The Top-Secret Food That Will Change the Way You Eat." *Outside Online*, June 30, 2021. https://www.outsideonline.com/culture/food/top-secret-food-will-change-way-you-eat/.

"Los Angeles Superior Court Rules in Favor of Beyond Meat on Don Lee Farms' Trade Secret Misappropriation and Unfair." *Bloomberg*, August 31, 2021. https://www.bloomberg.com/press-releases/2021-08-31/los-angeles-superior-court-rules-in-favor-of-beyond-meat-on-don-lee-farms-trade-secret-misappropriation-and-unfair.

Moskin, Julia. "How Do the New Plant-Based Burgers Stack up? We Taste-Tested Them." *The New York Times*, October 22, 2019. https://www.nytimes.com/2019/10/22/dining/veggie-burger-taste-test.html.

Starostinetskaya, Anna. "Beyond Meat Files for 100 Trademarks for Vegan Milk, Bacon, Eggs, and Seafood." *VegNews*, August 17, 2021. https://vegnews.com/2021/8/beyond-meat-trademarks.

Watson, Elaine. "Beyond Meat Founders: 'We're a Meat Company That Makes Products from Plants'." *Food Navigator USA*. William Reed Ltd, April 4, 2014. https://www.foodnavigator-usa.com/Article/2014/04/04/Beyond-Meat-We-re-a-meat-company-that-makes-products-from-plants.

Watson, Elaine. "The Top-Secret Food That Will Change the Way You Eat." *Outside Online*, June 30, 2021. https://www.outsideonline.com/culture/food/top-secret-food-will-change-way-you-eat/#close.

S. Canva

Benitez, Christopher. "11 Key Lessons from CANVA's Growth Strategy You Need to Know." *Bill Acholla*, September 19, 2021. https://www.billacholla.com/canvas-growth-strategy-lessons/.

Brand Minds. 2021. "Growth Story: How CANVA Acquired 10 Million Users within 5 Years." *Medium*. August 25. https://blog.markgrowth.com/growth-story-how-canva-acquired-10-million-users-within-5-years-bfe5275b321c#:~:text=Canva%20grew%20at%20a%20fast,our%20users%20sharing%20about%20us.

Canva Business Model Canvas. *Vizologi*. Accessed May 24, 2022. https://vizologi.com/business-strategy-canvas/canva-business-model-canvas/#home.

Canva. *Wikipedia*, May 18, 2022. https://en.wikipedia.org/wiki/Canva.

"How CANVA Design School Increased Traffic to 1 Million Visits - Learn." *Canva*. Accessed May 24, 2022. https://www.canva.com/learn/content-marketing-strategy/.

How Does Linkedin Make Money? 2021. *Investopedia*. June 9. https://www.investopedia.com/ask/answers/120214/how-does-linkedin-lnkd-make-money.asp.

Microsoft-LinkedIn Timeline: An inside Look at the Merger. 2021. *Wall Street Prep*. September 28. https://www.wallstreetprep.com/knowledge/a-look-at-the-microsoft-linkedin-merger/.

Mishra, Pankaj. "Guy Kawasaki Joins Australian Design Startup Canva as Chief Evangelist." *TechCrunch*, April 16, 2014. https://techcrunch.com/2014/04/16/guy-kawasaki-joins-australian-design-startup-canva-as-chief-evangelist/.

Sethi, Jayant. "How Canva's Strategy Is Challenging the Tech Giants." *The Strategy Story*, March 11, 2021. https://thestrategystory.com/2021/02/16/canva-freemium-business-model/.

Vivekanandarajah, An. "CANVA Marketing Strategy: How They Grew to 15 Million Users." *Hype Insight*, April 9, 2020. https://hypeinsight.com/canva-marketing-strategy-how-they-grew-to-15-million-users/.

T. MasterClass

Benson, Richard. "The Dream Academy: How Masterclass Became a Surprise Hit over Lockdown." *Esquire*, October 6, 2020. https://www.esquire.com/uk/culture/a34270753/what-is-masterclass/.

Friend, Tad. 2021. "Can Masterclass Teach You Everything?" *The New Yorker*. October 15. https://www.newyorker.com/magazine/2021/10/25/can-masterclass-teach-you-everything.

Masterclass. *Wikipedia*, February 28, 2022. https://en.wikipedia.org/wiki/MasterClass.

Nevins, Jake. "It's the Year 2120. Masterclass Is the Only School Left." *The New York Times*, May 25, 2020. https://www.nytimes.com/2020/05/25/style/masterclass-secrets.html.

U. Winterlight Labs

Holczinger, Andras and Liam Kaufmann. "Interview with Liam Kaufmannco-Founder & CEO of Winterlight Labs." *Global HealthTech Review*, n.d. https://www.healthtechbase.org/interview-2006-winterlight/.

Robin, Jessica, Mengdan Xu, Liam D. Kaufman, and William Simpson. 2021. "Using Digital Speech Assessments to Detect Early Signs of Cognitive Impairment." *Frontiers*. October 27. https://www.frontiersin.org/articles/10.3389/fdgth.2021.749758/full.

Winterlight Labs, n.d. "Winterligh Labs: Monitoring Cognitive Impairment Through Speech." https://winterlightlabs.com/.

V. Blue Rock Therapeutics

About. 2021. *Blue Rock Therapeutics*. July 30. https://bluerocktx.com/about/.

Approach. 2022. *Blue Rock Therapeutics*. April 12. https://bluerocktx.com/approach/.

"Bayer Acquires BlueRock Therapeutics." 2019. *University Health Network*. August 12. https://www.uhn.ca/corporate/News/Pages/Bayer_Acquires_BlueRock_Therapeutics.aspx.

Bluerock Therapeutics. 2022. *Engineered Cell Therapy*. April 12. https://bluerocktx.com/.

BlueRock Therapeutics Receives Permission from Health Canada for DA01 Trial in Parkinson's Disease. 2021. *Engineered Cell Therapy*. April 7. https://bluerocktx.com/bluerock-therapeutics-receives-permission-from-health-canada-for-da01-trial-in-parkinsons-disease/.

"Gordon Keller, Phd." *University Health Network Research*. https://www.uhnresearch.ca/researcher/gordon-keller.

"Gordon M. Keller." 2022. *Wikipedia*. January 21. https://en.wikipedia.org/wiki/Gordon_M._Keller.

Versant Ventures Announces Acquisition of Cell Therapy Company. 2019. *Versant Ventures*. August 8. https://www.versantventures.com/system/uploads/fae/file/asset/67/Versant_Ventures_Announces_Acquisition_of_Cell_Therapy_Company_BlueRock_Therapeutics___Business_Wire.pdf.

W. Braze Mobility

Beaulieu, Liz. 2021. "Braze Mobility Takes 'Next Step' with Va." *HME News*. November 23. http://www.hmenews.com/article/braze-mobility-takes-next-step-with-va.

Braze Mobility: Blind Spot Sensors for Wheelchairs. 2021. *Braze Mobility*. August 13. https://brazemobility.com/.

PIRL Webinar Dr Pooja Feb 14 Faster and BRAZE. 2022. *PIRL Project*. YouTube. https://www.youtube.com/watch?v=j9WAXrfuR0E.

X. Nuula

About Us - Nuula: Your Business at Your Fingertips. *Nuula*. https://nuula.com/about-us.

Canadian Startup Nuula Secures US$120 Million to Boost Service Offerings for SMEs. 2021. *Fintech Schweiz Digital Finance News*. September 15. https://fintechnews.ch/canada/canadian-startup-nuula-secures-us120-million-to-boost-service-offerings-for-smes/48850/.

Ruddock, Mark. "About Me." *Mark Ruddock*. https://markruddock.com/about-me.

Y. Drisit

How It Works. *Drisit*. https://drisit.com/how-it-works-1.

Z. Wordle

Kooser, Amanda. "Another Wordle Archive Shuts down under New York Times Pressure." *CNET*, April 13, 2022. https://www.cnet.com/culture/internet/how-you-can-play-every-wordle-ever-made/.

Lunden, Ingrid, and Amanda Siberling. "Wordle Founder Josh Wardle on Going Viral and What Comes Next." *Yahoo! News*. Accessed May 24, 2022. https://ca.news.yahoo.com/wordle-tech-born-love-asks-175205738.html.

Reeve, Justin. "The New York Times Acquires Wordle." *TheGamer*, January 31, 2022. https://www.thegamer.com/new-york-times-acquires-wordle/.

"The New York Times Company Reports 2022 First-Quarter Results." 2022. *The New York Times Company*. May 4. https://investors.nytco.com/news-and-events/press-releases/news-details/2022/The-New-York-Times-Company-Reports-First-Quarter-2022-Results/default.aspx.

Vats, Varsha. "Before Selling Viral Game Wordle, Founder Josh Wardle Chose to Pay Money for It." *DailyO*. Living Media India Limited, February 2, 2022. https://www.dailyo.in/variety/wordle-game-new-york-times/story/1/35314.html.

Victor, Daniel. "Wordle Is a Love Story." *The New York Times*, January 3, 2022. https://www.nytimes.com/2022/01/03/technology/wordle-word-game-creator.html.

"Wordle." *Wikipedia*, May 24, 2022. https://en.wikipedia.org/wiki/Wordle.

Further Reading

Brandt, Steven C. 1997. *Entrepreneuring: The 10 Commandments for Building a Growth Company*. Friday Harbor, Washington: Archipelago Publishing.

Chestnut, Robert, and Joan O'C. Hamilton. 2020. *Intentional Integrity: How Smart Companies Can Lead an Ethical Revolution*. Macmillan.

Christensen, Clayton M. 2000. *The Innovator's Dilemma: The Revolutionary Book That Will Change the Way You Do Business*. New York City, New York: Harper Business.

Crutchfield, Leslie R., and Heather McLeod Grant. 2012. *Forces for Good: The Six Practices of High-Impact Nonprofits*. San Francisco, Califronia: Jossey-Bass.

Cusumano, Michael A. 2004. *The Business of Software: What Every Manager, Programmer, and Entrepreneur Must Know to Succeed in Good Times and Bad*. New York City, New York: Free Press.

DiBona, Chris, Sam Ockman, and Mark Stone. 1999. *Open Sources: Voices from the Open Sources Revolution*. Beijing: O'Reilly.

Kanter, Beth, Allison H. Fine, and Randi Zuckerberg. 2010. *The Networked Nonprofit: Connecting with Social Media to Drive Change*. John Wiley and Sons.

Kawasaki, Guy. 2004. *The Art of the Start: The Time-Tested, Battle-Hardened Guide for Anyone Starting Anything*. New York City, New York: Portfolio.

Kennedy, Wendy, Peter Eddison, and John Hanak. 2015b. *So What? Who Cares? Why You? A Methodology to Find the Business Value of New Ideas*. Ottawa, Ontario: wendykennedy.com Inc.

Livingston, Jessica. 2008. *Founders at Work: Stories of Startups' Early Days*. Berkeley, California: APress.

Moore, Geoffrey A. 1991. *Crossing the Chasm: Marketing and Selling Technology Products to Mainstream Customers*. New York City, New York: Harper Business.

Nesheim, John L. 2000b. *High Tech Start Up: The Complete Handbook for Creating Successful New High Tech Companie*. New York City, New York: Free.

Osterwalder, Alexander, and Yves Pigneur. 2013. *Business Model Generation A Handbook for Visionaries, Game Changers, and Challengers*. New York City, New York: Wiley & Sons.

Rich, Stanley R., and David E. Gumpert. 1985. *Business Plans That Win $$$: Lessons from the MIT Enterprise Forum*. New York City, New York: Harper & Row.

Ries, Eric. 2011. *The Lean Startup: How Today's Entrepreneurs Use Continuous Innovation to Create Radically Successful Businesses*. Crown Business.

Suber, Peter. 2012. *Open Access*. Cambridge, Massachusetts: MIT Press.

Weber, Steve. 2005. *The Success of Open Source*. Cambridge, Massachusetts: Harvard University Press.

Index

Printed in the United States
by Baker & Taylor Publisher Services